Springer Theses

Recognizing Outstanding Ph.D. Research

For further volumes:
http://www.springer.com/series/8790

Aims and Scope

The series "Springer Theses" brings together a selection of the very best Ph.D. theses from around the world and across the physical sciences. Nominated and endorsed by two recognized specialists, each published volume has been selected for its scientific excellence and the high impact of its contents for the pertinent field of research. For greater accessibility to non-specialists, the published versions include an extended introduction, as well as a foreword by the student's supervisor explaining the special relevance of the work for the field. As a whole, the series will provide a valuable resource both for newcomers to the research fields described, and for other scientists seeking detailed background information on special questions. Finally, it provides an accredited documentation of the valuable contributions made by today's younger generation of scientists.

Theses are accepted into the series by invited nomination only and must fulfill all of the following criteria

- They must be written in good English.
- The topic should fall within the confines of Chemistry, Physics, Earth Sciences, Engineering and related interdisciplinary fields such as Materials, Nanoscience, Chemical Engineering, Complex Systems and Biophysics.
- The work reported in the thesis must represent a significant scientific advance.
- If the thesis includes previously published material, permission to reproduce this must be gained from the respective copyright holder.
- They must have been examined and passed during the 12 months prior to nomination.
- Each thesis should include a foreword by the supervisor outlining the significance of its content.
- The theses should have a clearly defined structure including an introduction accessible to scientists not expert in that particular field.

Addisson Salazar

On Statistical Pattern Recognition in Independent Component Analysis Mixture Modelling

Doctoral Thesis accepted by
Polytechnic University of Valencia, Spain

Springer

Author
Dr. Addisson Salazar
Department of Communications
School of Telecommunication Engineering
Polytechnic University of Valencia
Camino de Vera s/n
46022 Valencia
Spain

Supervisors
Prof. Luis Vergara
Department of Communications
School of Telecommunication Engineering
Polytechnic University of Valencia
Camino de Vera s/n
46022 Valencia
Spain

Prof. Jorge Igual
Department of Communications
School of Telecommunication Engineering
Polytechnic University of Valencia
Camino de Vera s/n
46022 Valencia
Spain

ISSN 2190-5053 ISSN 2190-5061 (electronic)
ISBN 978-3-642-30751-5 ISBN 978-3-642-30752-2 (eBook)
DOI 10.1007/978-3-642-30752-2
Springer Heidelberg New York Dordrecht London

Library of Congress Control Number: 2012941630

Printed on acid-free paper

Springer is part of Springer Science+Business Media (www.springer.com)

Parts of this thesis have been published in the following journal articles and patent:

Salazar A., Vergara L., Serrano A., Igual J., A General Procedure for Learning Mixtures of Independent Component Analyzers, Pattern Recognition, vol. 43 no. 1, pp. 69–85, 2010.

Salazar A., Vergara L., Miralles R., On including sequential dependence in ICA mixture models, Signal Processing, vol. 90, pp. 2314–2318, 2010.

Salazar A., Vergara L., Llinares R., Learning Material Defect Patterns by Separating Mixtures of Independent Component Analyzers from NDT Sonic Signals, Mechanical Systems and Signal Processing, vol. 24 no. 6, pp. 1870–1886, 2010.

Salazar A., Vergara L., ICA Mixtures Applied to Ultrasonic Non-destructive Classification of Archaeological Ceramics, Journal on Advances in Signal Processing, vol. 2010, Article ID 125201, 11 pages, doi:10.1155/2010/125201, 2010.

Salazar A., Vergara L., Gosalbez J., Miralles R., Bosch I., Parra A., "Unidad y Metodo de Catalogación de Ceramicas Arqueologicas segun su Procedencia" (Unit and method to catalogue archaeological ceramics according to provenance), Spanish Office of Patents and Trademarks, 2010.

In the sphere of thought, sober civilization is roughly synonymous with science. But science, unadulterated, is not satisfying; men need also passion and art and religion. Science may set limits to knowledge, but should not set limits to imagination.

Bertrand Russell, *History of Western Philosophy*—The Pre-Socratics, The Rise of Greek Civilization (George Allen & Unwin Ltd, London, 1946)

To Nancy

Supervisor's Foreword

The concrete objective of this thesis is the extraction of information and patterns on an underlying process from a number of observations. This task becomes increasingly difficult without any prior knowledge, e.g. from a physical analysis of the underlying process, or from prior training. Overcoming these difficulties step by step is the main objective of this thesis. The systematic approach developed and the consequent application of the theoretical results to a large number of practical problems clearly demonstrates the great potential interest of this work. The focus of the methods described is on defining a general framework in statistical pattern recognition based on independent component analysis mixture modelling (ICAMM).

The first core step is the incorporation of non-parametric estimation for the a priori distributions in statistical pattern recognition problems, considering that the features are vectors resulting from a mixture of hidden independent random variables. Keeping in mind this purpose, additional relevant problems are approached: correction of the posterior distributions; completely supervised and partially supervised scenarios considering the impact of the unsupervised features; fitting of the corresponding learning methods in mixture matrix/centre of gravity; and a detailed analysis of all the ICAMM steps. The second core step is to develop a clustering method that allows the number of classes to be determined from the features extracted by the ICAMM analysis. The novelty consists in evaluating the distances between clusters not by using the features that make it up, but based on its distributions. The method to estimate both the entropy and the cross entropy in the cluster structure using non-parametric methods is provided. Therefore, the formalisms introduced in the thesis unify a certain number of pattern recognition tasks achieving generalisation. The suitability for the proposed objectives is demonstrated at first by a large number of simulations with synthetic data to show the advantages of the introduced algorithms. Second, various applications from quite different fields are approached: material quality control using the impact-echo technique; chronological cataloguing of archaeological ceramics; object recognition and image segmentation; diagnosis of historic building restoration; diagnosis of sleep disorders; and the discovery of learning styles in e-learning.

The manuscript itself is a well-organised piece of work with an excellent state-of-the-art literature review, which allows the contributions of the thesis to be precisely realised. A consistent methodology is developed based on independent component analysis and extends the initial approach by considering mixtures of different linear models and by considering nonlinear elements. The theoretical extensions are thoroughly explained and emphasised through appropriate and demonstrative examples. This provides a comprehensive understanding of the application of the methods to solve diverse real-world problems. The contributions of the thesis are perfectly closed and have been worthy of publication in outstanding journals, and from them, a patent in the field of archaeological cataloguing has been invented. Also, several future lines of research are proposed from this work, such as: residual dependence after convergence in hierarchical clustering, incorporation of the use of priors in the estimation of the source densities, development of techniques to detect and process outliers, extension to sequential methods, and incorporation of online estimation of the parameters.

The achieved results are without doubt of interest in a wide variety of application fields. These results demonstrate the capability and versatility of the proposed methods to be adapted to different problems in order to find significant structures in data. In most practical cases, desired information can only be accessed indirectly due to mixing of several sources of information, introducing of bias, nonlinear mappings and noise. Conventional methods have to rely on assumptions, like Gaussian distributions, linearity or the validity of parametric models. These restrictions can be overcome by the framework discussed in this thesis.

Valencia, 08 April 2012

Dr. Luis Vergara
On behalf of Prof. Jorge Igual

Acknowledgments

First and foremost, I must thank my first supervisor Luis Vergara for the invaluable guidance, encouragement and inspiration that he has given me over the course of my studies. He has generously shared his expertise and research insight with me and has supported me to broaden my research objectives.

I must also thank my second supervisor Jorge Igual for introducing the subject of this thesis to me, for conveying his research enthusiasm, and for his enlightening discussions, all of which have helped me gain a deeper understanding of my research.

I would like to thank Vicente Zarzoso, who allowed me to stay at his lab at the Université Nice Sophia Antipolis, and for his valuable comments about this work.

Thanks are also in order for my colleagues at the Signal Processing Group (GTS), Jorge Gosálbez, Ignacio Bosch and Ramón Miralles. I have enjoyed the time that we have shared in planning, designing and developing diverse research projects. It has been a truly enriching professional and human experience.

I would also like to thank my sisters Irma and Olinda, my brother Pedro and my niece Maria Paula for their encouragement and love over the years.

Finally, I would like to express my gratitude to the many people who have made this thesis possible at GTS, Institute of Telecommunications and Multimedia Applications (iTEAM), and Department of Communications at Polytechnic University of Valencia.

Mathematical Notations

Independent component analysis mixture modelling

K	Number of classes (ICA models), $k = 1,\ldots,K$
M	Number of sources (dimensionality), $m = 1,\ldots,K$
N	Number of observed-mixture vectors, $\mathbf{X} = [\mathbf{x}^{(1)} \cdots \mathbf{x}^{(N)}]$
$x_{ki}^{(n)}, \mathbf{x}_k^{(n)}$	i th sample of the n th observed-mixture vector of the k th class, n th observed- mixture vector of the k th class
$s_{km}^{(n)}, \mathbf{s}_{km}$	n th sample of the m th source vector of the k th class, m th source vector of the k th class
$\mathbf{A}_k$	mixing matrix of the k th class
$\mathbf{W}_k$	demixing matrix of the k th class
$\mathbf{b}_k$	centroid of the k th class
$\hat{\mathrm{S}}$	Separability matrix

Statistics and mathematics

$p(\cdot)$	Probability
$p(A/B)$	Conditional probability of A, given B
$p(A, B)$	Joint probability of A, given B
$D_{KL}(p, q)$	Symmetric Kullback-Leibler distance between densities p and q
$\det \mathbf{A}$	Determinant of matrix $\mathbf{A}$
σ, σ^2	Standard deviation, variance
$(\cdot)^{-1}$	Inverse
$\hat{(\cdot)}$	Estimate
$(\cdot)^T$	Transpose
$\vert \cdot \vert$	Absolute value
$tr\,[\cdot]$	Trace of a matrix
$E\,[\cdot]$	Mathematical expectation
$diag\,[\cdot]$	Transforms a vector into a diagonal matrix
$L\,(\cdot)$	log-likelihood
$\overline{\max}(\cdot)$	Mean of the maximum

cov $(\cdot)$	Covariance
$\log A$	Logarithm of A
δ	Partial derivate operator
Δ	Increment, gradient operator
$\Leftrightarrow$	If, and only if

Signal processing

$x(t)$	Time-domain signal
$\mathcal{F}\{\cdot\}, \mathcal{F}^{-1}\{\cdot\}$	Fourier transform, inverse Fourier transform
$BPF\ (f \pm \Delta)$	Narrow band pass filter centred in f
$H_{ij}(\omega)$	Frequency response (transfer function) between j and i

Acronyms

AR	Autoregressive
ASI	Alpha slow-wave index
BSE	Blind signal extraction
BSS	Blind source separation
CCA	Canonical correlation analysis
COIL	Columbia object image library
DIL	Blind signal extraction
DFT	Blind source separation
DPL	Distributed passive learning
ECG	Electrocardiogram
EEG	Electroencephalogram
EM	Expectation maximization
FastIca	FAST fixed-point algorithm for independent component analysis
FEM	Finite element method
FFT	Fast fourier transform
fMRI	Functional magnetic resonance imaging
HMM	Hidden markov model
HOS	Higher-order statistics
ICA	Independent component analysis
ICAMM	Independent component analysis mixture modelling
i.i.d.	Independent and identically distributed
InfoMax	INFOrmation MAXimization
JADE	Joint approximate diagonalization of eigen-matrices
Kernel-ICA	Kernel independent component analysis
KL	Kullback–Leibler
kNN	k-Nearest neighbours
LDA	Linear discriminant analysis
LTI	Linear time invariant
LTV	Linear time varying
LVQ	Learning vector quantization
MAP	Maximum A posteriori estimation

MIMO-LTI	Multiple input multiple output-LTI
Mixca	MIXture of component analyzers (non-parametric density estimation)
Mixca-Alg	MIXCA variant (Alg: parameter updating algorithm,e.g., JADE, TDSEP)
MLE	Maximum likelihood estimation
MLP	Multi-layer perceptron
MoG	Mixture of Gaussians
NDT	Non-destructive testing
NP	Non-parametric
Npica	Non-parametric ICA
PC	Partition coefficient
PCA	Principal component analysis
pdf	Probability ensity function
PE	Partition entropy coefficient
PPCA	Probabilistic principal component analysis
PSG	Polysomnogram
PSS	Probabilistic semi-supervision
QBSE	Query by semantic example
QBVE	Query by visual example
Radical	Robust, Accurate, Direct ICA aLgorithm
RBF	Radial basis function
RDS	Residual deconvolutioned signal
REM	Rapid-eye movement
SBSS	Semi-blind source separation
SEM	Scanning electron microscope
SICAMM	Sequential ICAMM
SIR	Signal to interference ratio
SNR	Signal to noise ratio
SSL	Semi-supervised learning
sr	Supervision ratio
TCA	Tree-dependent component analysis
TDSEP	Temporal decorrelation source separation
TICA	Topographic independent component analysis
TSI	Theta slow-wave index
w.r.t.	With regard to

Contents

Chapter 1
Introduction

From our most early experiences with reality, we start to recognize patterns in the surrounding environment. This allows us as human beings to be aware of the different objects that we are related to. The scope of pattern recognition is broad since it is observed at different levels in the world. This awareness occurs for a cell that divides and specializes itself and for an expert standing in front of a painting trying to make a distinction between the pure object and the pure subject of that object. This natural skill of living, which is the basis of the learning process, is artificially imitated by means of sequences of mathematic-computational steps known as *machine learning*. Without going into an epistemological discussion about whether or not the perceived reality is biased by our senses and capabilities to process what is perceived, the high complexity of the tasks for machine learning represents an actual challenge. The questions of why and what we learn for, which are generally related to adaptation for evolving, are outside the scope of this work. Instead, we will focus on how to learn artificially, and we will propose mathematical procedures that are able to distinguish, represent, and allocate learning objects, as well as assess these procedures in novel application fields.

Learning can be defined as the process of inferring general rules from given examples. To obtain an artificial approach to the learning process, we have to simplify the representation of the objects of the real-world whose patterns must be recognized and/or learned. Afterwards, a way to search for data of interest in these objects must be proposed. The data of interest will usually have (or we attempt to find) a certain meaning or interpretation that allocates them to a specific field of application. Thus, real objects or phenomena are represented as numbers or data, i.e., collections of ones and zeros for computer processing. Data are obtained from measurements applied on the real objects by means of sensors. The estimation and analysis of the distributions and groupings as well as the inference on possible former generators of these data are critical for pattern recognition.

There are several definitions that have been proposed for pattern recognition [1]. In this thesis, we will assume a common accepted explanation of pattern

A. Salazar, *On Statistical Pattern Recognition in Independent Component Analysis Mixture Modelling*, Springer Theses, DOI: 10.1007/978-3-642-30752-2_1,

recognition as *a search for structure in data* [2]. The principal task of pattern recognition that we are interested in is in dividing a manifold of data into categories that have a meaning under an application context. We will approach this task from two perspectives: classification and clustering. The first one considers a predefined set of categories to which data items can be assigned; the second one discovers significant groups that are present in a data set with no predefined classes and no examples that would show what kind of desirable relations should be valid among the data. The structures searched for in the data consist of the rules that describe the relationships in the data. These structures can be defined through a probabilistic model that can provide a reasonable explanation of the process generating the data. We assume that the set of observed variables that are explicitly defined in the data are generated from a set of hidden variables of an underlying model. Thus, the data are denoted by formulae or models that describe their principal characteristics. The ratio of the complexity of the data set to the complexity of the formulae is defined as *parsimony*. In order to propose a model for the data, it is necessarily assumed that there exist patterns or rules in the data. In this case, the data are redundant, and the patterns may be used to provide a parsimonious description that is more concise than the data themselves [3].

Pattern recognition is frequently achieved by using features extracted from the raw data either because the stream of the measured data is large or because processing raw data does not allow patterns to be distinguished. Thus, the selection and estimation of the features should lead to adequately characterizing the conspicuous properties of the data. An appropriate set of features allows the data to be separated in different groups or clusters, where the data in one group are the most similar to each other, and are also the most dissimilar to the data in others groups. The groups of data extracted from the original data manifold represent particular patterns with particular meanings for which an explicit label could be assigned. Once the rules for the patterns are learned from a dataset, they can be applied to classify new datasets.

There can be different degrees of completeness of the knowledge of the labels for the dataset employed in the learning process. The degree of knowledge available determines the kind of learning, i.e., supervised (all the data-label pairs are known), semi-supervised (labels are available for a subset of the data), and unsupervised (no labels are available). The kind of learning must be encompassed within the complexity of the real-world problem that imposes a minimum level of labelling in order to learn the geometry of the data. Increasing the data labels available in order to reach an adequate level is restricted for several applications. Frequently, obtaining unlabelled data may be relatively easy whereas obtaining labelled data may be difficult and costly. However, the latter can be alleviated by considering that the performance of some algorithms is significantly improved with a small number of labelled data [4, 5]. Therefore, semi-supervised learning has been increasingly studied (for its capability to incorporate different proportions of unlabelled and labelled data) as a suitable method for many complex problems [6].

Intelligent signal processing algorithms provide an important tool to support automatic pattern recognition, to gain insight into problem-solving, and to complement expert decision-making 0 [7–9]. These algorithms usually make assumptions about the data-generating process, for instance, modelling the data as a mixture of data generators or analyzers considering that each generator produces a particular group of data. The independent component analysis mixture modelling (ICAMM) [10, 11] has recently emerged as a flexible approach to model arbitrary data densities using mixtures of multiple independent component analysis (ICA) models [12–14] with non-gaussian distributions for the independent components (i.e., relaxing the restriction of modelling every component by a multivariate Gaussian probability density function). ICA is an intensive area of research that is progressively finding more applications for both blind source separation (BSS) and for feature extraction/modelling. The goal of ICA is to perform a linear transformation of the observed sensor signals, such that the resulting transformed signals (the sources or prior generators of the observed data) are as statistically independent of each other as possible. In comparison to correlation-based transformations such as principal component analysis (PCA), ICA not only decorrelates the sensor observations composed of mixed signals (in terms of second-order statistics), but it also reduces the higher-order statistical dependencies among them [13–18].

Applications of ICA comprise such diverse disciplines as: speech separation; biomedical applications (removing electrocardiogram (ECG) and electroencephalogram (EEG) artefacts, noninvasive fetal ECG extraction, separation and determination of brain activity sources, diagnosis of atrial fibrillation, extraction of sources of neural activity in the brain) [19–23]; image processing (recognition of faces using ICA bases, spatial edge filters using separating matrix, reconstruction and restoration of distorted images); text classification; non-destructive testing (NDT) (analysis of the vibration in mechanical systems: termite activity in wood, identification of transient low-events in diesel engines, identification of particular faults for gearbox diagnostics); and telecommunications (remove interfering transmission in wireless telecommunications systems, blind code-division multiple access) [13–18].

The linear ICA method is extended in ICAMM to a kind of nonlinear ICA model, i.e., multiple ICA models are learned and weighted in a probabilistic manner. Thus, the ICA mixture model is a conditional independence model, i.e., the independence assumption holds only within each class and there may be dependencies among the classes [10]. The degrees of freedom afforded by mixtures of ICAs allow a broad range of real problems involving complex data densities to be dealt with. ICAMM contributes to obtaining higher insights into the applications since this modelling performs both source extraction and signal analysis simultaneously. This enables a more detailed explanation of the measured signals and of the source data generators that are behind the observed mixture. The suitability of mixtures of ICA for a given problem of data analysis and classification can be treated from different perspectives. First, there is the "least physical" interpretation, which assumes that ICA mixture learning underlies estimation/modelling of the probability density of multivariate data [11]. Second, there is the interpretation of ICA as a way of learning some bases (usually called activation functions), which are more or less connected to the actual behaviours that are implicit in the physical phenomenon under analysis [24]. Third, there is the "most physical" interpretation, which attempts to identify where sources are originated and how they mix before arriving to the sensors to provide a physical explanation of the linear mixture model. In any case, even though the complexity of the problem constrains a physical interpretation, ICAMM can be used as a general-purpose data mining technique.

ICAMM has been applied in real applications such as: learning of natural image codes [10]; image classification, segmentation and denoising [25]; separation of voices and background music in conversations [10, 26]; unsupervised object image classification from multispectral and hyperspectral sensors [27, 28]; and separating several sources with fewer sensors in teleconferencing [29]. ICAMM has also been applied in biosignal processing: separation of background brain tissue, fluids and tumors in functional magnetic resonance imaging (fMRI) [11]; analysis to identify patterns of glaucomatous visual field defects [30, 31]; assessment of EEG to detect changes in dynamic brain state [32]; classification of breast cancer diagnostic data [33]; and analysis of multi-phase abdominal CT images to highlight liver segments [34].

The thesis explores statistical pattern recognition from the perspective of the mixtures of independent component analyzers. We will mainly be concerned with two problems: signal classification and BSS. Thus, a new approach for ICAMM that pursues generalization of this framework is proposed. The utility of the proposed methods is demonstrated in classic applications as well as in innovative applications that have not yet been attempted using ICA or using ICAMM. In this work, we develop several theoretical concepts that require a basic working-knowledge of Calculus and Probability. Thus, to make the understanding of these developments more accessible to the reader, a review of the main applied theoretical foundations of statistical pattern recognition is included in this chapter. The importance that these theories have for this work is also discussed. The rest of the Introduction includes the scope and contributions addressed in the thesis and an overview of the rest of the chapters.

1.1 Statistical Pattern Recognition

There are three basic aspects to be taken into account in the design of a pattern recognition system: (i) data acquisition and preprocessing; (ii) data representation; and (iii) decision making. These aspects are conditioned by the application domain. The four best known approaches for pattern recognition are: (i) template matching, (ii) statistical classification, (iii) syntactic or structural matching, and (iv) neural networks [8]. The procedures of this work are proposed from the statistical classification approach. Statistical pattern recognition is based on the statistical decision theoretical approach. This approach assumes that the data of different classes can be separated according to decision boundaries determined by the probability distributions of the data. Normally, the raw data are projected in a

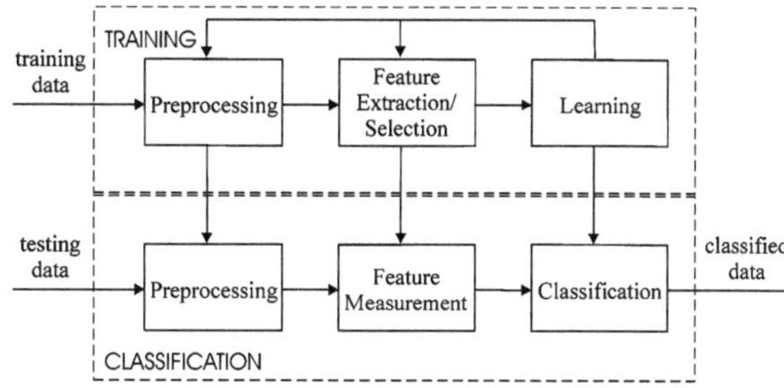

Fig. 1.1 Statistical pattern recognition process

d-dimensional space defined by d selected features (transforms of the data). Data classification is carried out in the feature space. Thus, a feature set establishes regions (decision boundaries) with a degree of overlapping in order to separate the data belonging to different classes. The efficiency of the feature set is the capability to establish compact and disjoint regions. The recognition process has two stages: training (decision boundaries are learned from the training data) and classification (new data are presented for testing), see Fig. 1.1.

The preprocessing task consists of the operations for obtaining a clean and compact representation of the pattern (removing noise, normalization, etc.). The feature extraction and selection task provides an adequate set of features to the learning task. The algorithms used for feature ranking are intended to estimate the contribution of each feature for a correct assigning of the classes to the training data. Feature selection is a classic problem in pattern recognition, and many methods have been proposed to solve it. These methods are based on information gain, margins with respect to a classification rule, weight-based methods, etc. [35]. Another issue to solve in this task is reduction of dimensionality when the number of features is large. One of the most popular methods for dimensionality reduction is PCA, which maps the features to a new space of components that are linear combinations of the original features [13]. The final feature set extracted from the training data is used for the learning algorithm to estimate the decision boundaries, i.e., the partition of the feature space. Figure 1.2 shows a scatter plot of a 2-dimensional, 2-class dataset of 200 objects. Features are Gaussian distributions, and classes are equally probable. Different shapes of decision boundaries that partition the feature space obtained by different classification algorithms are depicted.

The training stage can be iterative, and feedback is usually required from the learning task to the previous tasks in order to tune the training to the data. In the classification stage, the trained classifier assigns each of the input testing data to one of the classes under consideration based on the calculated features. In the decision making process, the features are assumed to have a probability density function (pdf) conditioned on the data class. Thus, a feature data vector $\mathbf{x} = (x_1, x_2, \ldots, x_d)$ is viewed as an observation drawn randomly from the class-conditional probability function $p(\mathbf{x}|C_k)$, considering $k = 1, \ldots, K$ classes.

Various approaches have been utilized for classifier design in statistical pattern recognition. They can be organized depending on the kind of information available

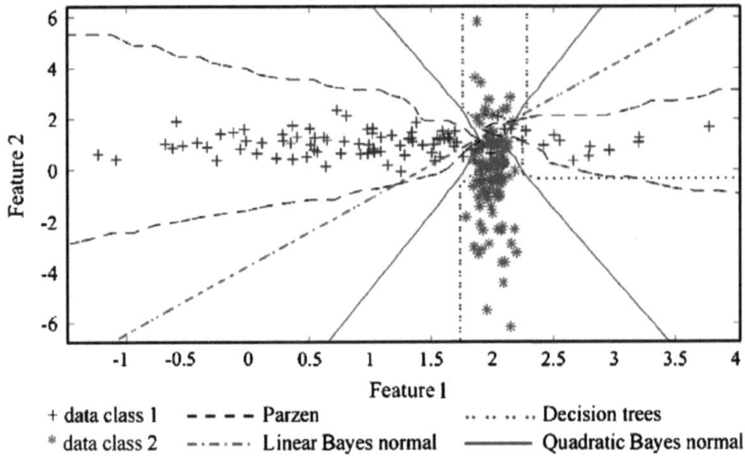

Fig. 1.2 Decision boundaries established by different classifiers

about the class-conditional densities. Figure 1.3 shows a summary of the statistical classifiers that can be found in the literature [8] 0, highlighting the types of classifiers that are proposed in this work. Statistical classifiers can be summarized in three categories: (i) based on the concept of similarity—defining an appropriate distance metric; (ii) based on the probabilistic approach; the optimal Bayes decision rule (with 0/1 loss function) assigns a pattern to the class with the maximum posterior probability; and (iii) based on the construction of decision boundaries (geometric approach) directly by optimizing certain error criterion.

There are a number of decision rules available to define the decision boundaries, for instance, Bayes decision, maximum likelihood, and Neyman-Pearson. The decision rule that attempts to implement most of the statistical classifiers, including the ones proposed in this work, is the Bayes decision rule. The "optimal" Bayes decision rule is stated to minimize the conditional risk $R(C_i|\mathbf{x})$ of assigning input data $\mathbf{x}$ to class C_i. Thus,

$$R(C_i|\mathbf{x}) = \sum_{j=1}^{K} L(C_i, C_j) \cdot P(C_j|\mathbf{x}) \tag{1.1}$$

where $L(C_i, C_j)$ is the loss incurred in deciding C_i when the true class is C_j and $P(C_j|\mathbf{x})$ is the posterior probability 0 [36]. Assuming a 0/1 loss function, i.e., $L = 0,\ i = j$ and $L = 1,\ i \neq j$, the conditional risk becomes the conditional probability of misclassification and thus the objective is to minimize the probability of classification error. In this case, the Bayes decision rule is called the maximum a posteriori (MAP) rule and can be defined as follows: assign input data $\mathbf{x}$ to class C_i if

$$P(C_i|\mathbf{x}) > P(C_j|\mathbf{x}) \text{ for all } j \neq i \tag{1.2}$$

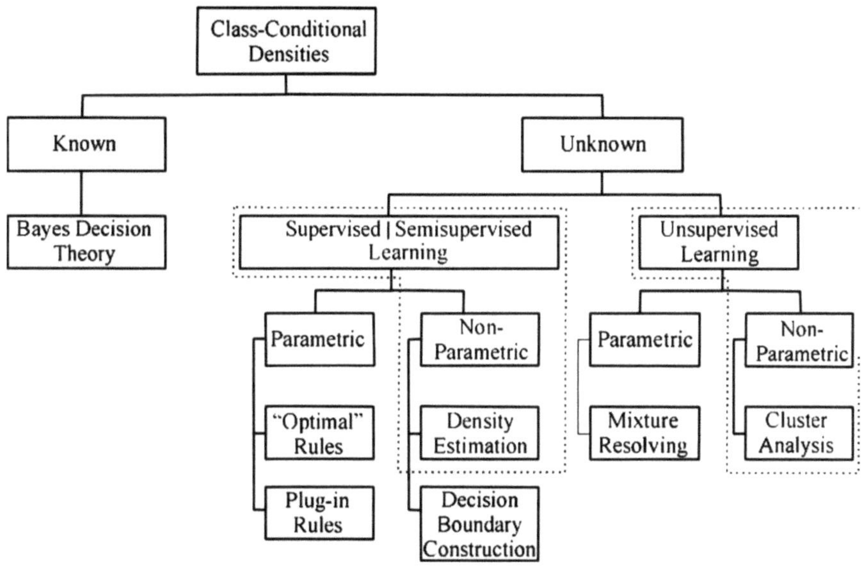

Fig. 1.3 Types of statistical classifiers. The types of methods proposed in this work are highlighted

1.1.1 Density Estimation

The probability density function is a fundamental concept in statistics. The probability density function f for a random quantity X gives a natural description of the distribution of X and also allows probabilities associated with X to be found from the relation

$$P(a < X < b) = \int_a^b f(x)dx \quad \text{for all } a < b \tag{1.3}$$

Suppose that we have a set of observed data points assumed to be samples from an unknown probability density function. Density estimation is the construction of an estimate of the density function from the observed data. One approach to density estimation is *parametric*, which assumes that the data are drawn from one of a known parametric family of distributions (e.g., the normal distribution with mean μ and variance σ^2). The density f underlying the data could then be estimated by finding estimates of μ and σ^2 from the data and substituting these estimates into the formula for the normal density. The *non-parametric* approach imposes less rigid assumptions about the distribution of the observed data. Although it is assumed that the distribution has a probability density f, the data is allowed to "speak for themselves" in determining the estimate of f more than would be the case if f were constrained to fall in a given parametric family.

There are many methods for density estimation such as histograms; naive estimator; nearest neighbour; variable kernel; orthogonal series; maximum penalized likelihood; general weight function; and bounded domains and directional data 0 [37]. We will use a kernel non-parametric approach for density estimation in the development of the thesis, so a brief review of the kernel-based method is included in this section. The kernel estimator is defined as

$$\hat{f}(x) = \frac{1}{nh}\sum_{i=1}^{n} K\left(\frac{x - X_i}{h}\right) \tag{1.4}$$

where h is the window width (also called the smoothing parameter or bandwidth), n is the number of observations, and K is a weight kernel function that satisfies the condition $\int_{-\infty}^{\infty} K(x)\,dx = 1$. $\hat{f}$ is a probability density that will inherit all the continuity and differentiability properties of the kernel K, i.e., if K is the normal density function, then $\hat{f}$ will be a smooth curve having derivatives of all orders. Figure 1.4 shows an example of density estimation using the kernel method of Eq. (1.4) for univariate data. The estimated density is superimposed on the histogram of the data.

A variable kernel is obtained by considering a scale parameter of the "bumps" that one placed on the data points, which is allowed to vary from one data point to another. The variable kernel estimate with smoothing parameter h is defined by

$$\hat{f}(t) = \frac{1}{n}\sum_{j=1}^{n} \frac{1}{hd_{j,k}} K\left(\frac{t - X_j}{hd_{j,k}}\right) \tag{1.5}$$

where $hd_{j,k}$ is the distance from X_j to the kth nearest point in the set comprising the other $n-1$ data points. The window width of the kernel placed on the point X_j is proportional to $d_{j,k}$ so that the data points in regions where the data are sparse will have flatter kernels associated with them. For any fixed k, the overall degree of smoothing will depend on the parameter h. The choice of k determines how responsive the window width choice will be to very local detail.

The quality of density estimate is evaluated by the closeness of the estimator $\hat{f}$ to the true density f. The estimate $\hat{f}$ depends on the data as well as on the kernel and the window width; this dependence will not generally be expressed explicitly. For each x, $\hat{f}(x)$ can be thought of as a random variable because of its dependence on the observations $X_1 \ldots, X_n$; any use of probability, expectation and variance involving $\hat{f}$ is with respect to its sampling distribution as a statistic based on these random observations. The analysis of statistical properties of the kernel estimator usually considers that the kernel K is a symmetric probability function satisfying $\int K(t)dt = 1$, $\int tK(t)dt = 0$, and $\int t^2K(t)dt = k_2 \neq 0$, and that the unknown density f has continuous derivatives of all orders required. In the case of a Gaussian kernel, k_2 will be the variance of the distribution with this density.

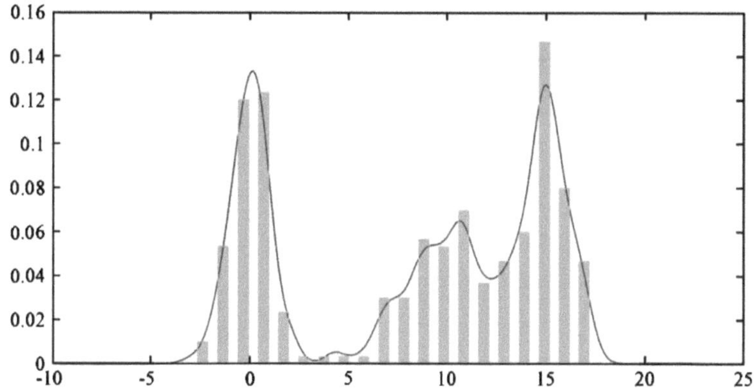

Fig. 1.4 Density estimation for univariate data

Table 1.1 shows the definition of some kernels used in non-parametric density estimation.

There are no significant differences among the various kernels to estimate an optimal window width (h_{opt}) on the basis of minimizing the approximate mean integrated square error. Thus, it is suggested to base the choice of the kernel, for example, on the degree of differentiability required or the computational burden involved. The problem of choosing how much smoothness is required is of crucial importance in density estimation. There are several methods proposed to estimate h_{opt}, for instance, subjective choice, reference to a standard distribution, least-squares cross-validation, likelihood cross-validation, the test graph method, and internal estimation of the density roughness 0 [37].

The kernel estimator as a sum of bumps centred at the observations for multivariate data is defined by the following expression

$$\hat{f}(\mathbf{x}) = \frac{1}{nh^d} \sum_{i=1}^{n} K\left\{ \frac{1}{h}(\mathbf{x} - \mathbf{X}_i) \right\} \tag{1.6}$$

The kernel function $K(\mathbf{x})$ is a function defined for d-dimensional $\mathbf{x}$, satisfying $\int_{R^d} K(\mathbf{x})\, d\mathbf{x} = 1$. K will be usually a radially symmetric unimodal probability density function, for example the standard multivariate normal density function

$$K(\mathbf{x}) = (2\pi)^{-d/2} e^{-\frac{1}{2}\mathbf{x}^T\mathbf{x}} \tag{1.7}$$

The use of a single smoothing parameter K in Eq. (1.6) implies that the version of the kernel placed on each data point is scaled equally in all directions. If the variance of the data points is very much higher in some of the coordinate directions and a pre-scale step is not done, a vector or matrix of smoothing parameters should be applied. However, if a pre-scale step is done, the Eq. (1.6) could be applied using a single smoothing parameter.

Table 1.1 Some kernel definitions

Kernel	$K(t)$
Gaussian	$\frac{1}{\sqrt{2\pi}}e^{-(1/2)t^2}$
Rectangular	$\frac{1}{2}$ for $\lvert t\rvert < 1$, 0 otherwise
Triangular	$1 - \lvert t\rvert$ for $\lvert t\rvert < 1$, 0 otherwise
Epanechnikov	$\frac{3}{4}\left(1 - \frac{1}{5}t^2\right)/\sqrt{5}$ for $\lvert t\rvert < \sqrt{5}$, 0 otherwise
Biweight	$\frac{15}{16}(1 - t^2)^2$ for $\lvert t\rvert < 1$, 0 otherwise

1.1.2 Semi-Supervised Learning

Traditionally, there are two different types of tasks in machine learning: supervised and unsupervised learning. For supervised learning, there is a sample $\{\mathbf{x}_i\}$ of patterns that are independently and identically distributed (i.i.d.) from some unknown data distribution with density $P(\mathbf{x})$ that has to be estimated. Supervised learning consists of estimating a functional relationship $\mathbf{x} \to \mathbf{y}$ between a covariate $\mathbf{x}$ and a class variable $y \in \{1, \ldots, M\}$, with the goal of minimizing a functional of the joint data distribution $P(\mathbf{x}, y)$ such as the probability of classification error. The marginal data distribution $P(\mathbf{x})$ is referred to as input distribution. Classification can be treated as a special case of estimating the joint density $P(\mathbf{x}, y)$.

Unsupervised learning can be considered as a density estimation technique. Many techniques for density estimation usually propose a latent (unobserved) class variable y and estimate $P(\mathbf{x})$ as mixture distribution $\sum_{i=1}^{M} P(\mathbf{x}|y)P(y)$. Note that the role of y in unsupervised learning is for modelling instead of being a role that is related to observable reality, which is the usual role of y in classification.

The semi-supervised learning (SSL) problem could be considered as belonging to any of the two previous categories of learning. If the goal is to minimize the classification error, semi-supervised learning would be a supervised task; however if the goal is to estimate $P(\mathbf{x})$, it would be an unsupervised task. In this latter case, the problem imposes more significance on the density estimation, and the labelled data are treated as an auxiliary resource. Thus, "semi-unsupervised learning" would be a more suitable name for this task. The difference with a standard classification setting is that along with a labelled sample $D_l = \{(\mathbf{x}_i, y_i)|i = 1, \ldots, n\}$ that is drawn i.i.d. from $P(\mathbf{x}, y)$ there is also has access to an additional unlabelled sample $D_u = \{(\mathbf{x}_{n+j}\,|j = 1, \ldots, m\}$ from the marginal $P(\mathbf{x})$. Of special interest are the cases where $m \gg n$, which may arise in situations where obtaining an unlabelled sample is cheap and easy, while labelling the sample is expensive or difficult. We denote $\mathbf{X}_l = (\mathbf{x}_1, \ldots, \mathbf{x}_n)$, $Y_l = (y_1, \ldots, y_n)$ and $\mathbf{X}_u = (\mathbf{x}_{n+1}, \ldots, \mathbf{x}_{n+m})$. The unobserved labels are denoted $Y_u = (y_{n+1}, \ldots, y_{n+m})$ [6].

SSL methods proceed in different ways, for instance, by first applying an unsupervised method for estimating mixture distribution $P(\mathbf{x})$, and then associating the latent groups obtained with observed classes using D_l. These methods are derived from one or more of the following assumptions: smoothness (the label function is smoother in high-density regions than in low-density regions); clustering (if points are in the same cluster, they are likely to be of the same class); manifold (the high-dimensional data lie roughly on a low-dimensional manifold; and transduction (directly estimating the finite set of test labels, i.e., $f : X_u \rightarrow Y$ only defined on the test set, instead of infers $f : X \rightarrow Y$ on the entire space X, and afterward returns $f(x_i)$ at the test points). Thus, SSL methods can be roughly organized into four classes depending on the assumptions considered: generative models (e.g., semi-supervised clustering with constraints, SSL using maximum likelihood estimation); low-density separation (e.g., transductive support vector machine, SSL using semi-definite programming, data-dependent regularization); graph-based methods (e.g., discrete regularization, SSL with conditional harmonic mixing); and change of representation (e.g., graph kernel by spectral transforms, spectral methods for dimensionality). Theoretical work in SSL has incorporated and adapted a diverse set of tools that were initially developed in other branches of machine learning such as kernel methods or Bayesian techniques. However, it has been stated that the relevant work of SSL is in practical subjects that are related to real-world applications [6]. Some examples of applications of SSL are classification of protein sequences, prediction of protein functions, speech recognition, and webpage classification. SSL only works in those cases where the knowledge on $p(x)$ gained through the unlabelled data carry information that is useful in the inference of $p(y|x)$. The existence of classes must also be guaranteed: if there is a densely populated continuum of objects, it may seem unlikely that they could ever be distinguished into different classes.

1.1.3 Hierarchical Clustering

There are many cases when a big cluster can be divided to meaningful subclusters, which can be divided into smaller subclusters, and so on. This kind of grouping procedure is called hierarchical clustering and is commonly used for summarising data structures. The most natural representation of hierarchical clustering is a corresponding tree, called a *dendrogram*, which shows how the samples are grouped, see Fig. 1.5. Clusters at a level are grouped depending on a similarity measure. This measure can be used to determine the best level of partition in the hierarchical structure, depending on the specifics of the application. Another representation for hierarchical clustering is using sets to represent the subclusters 0 [9].

Several similarity or distance measures have been proposed and these give rise to different hierarchical structures. Among the most common measures are: Euclidean, city-block, Chebyshev, Minkowski, quadratic, and Mahalanobis distance. These distances are defined between two data sets. There is another type of

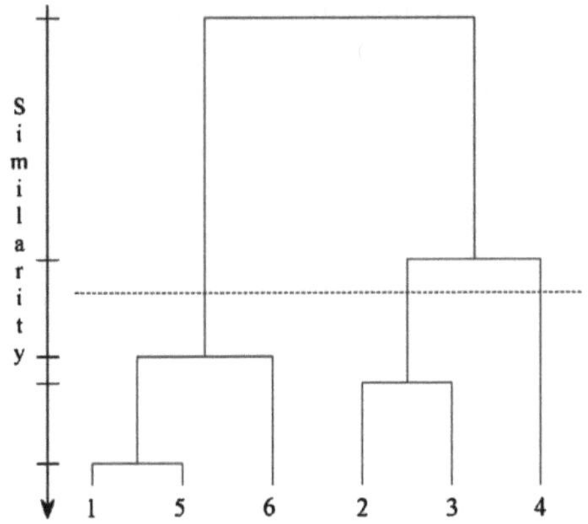

Fig. 1.5 A dendrogram example

distance that measures the distance between groups of data or data distributions. These probabilistic distance measures use the complete information about the structure of the classes provided by the conditional densities. Thus, the measure distance, J, satisfies the conditions: (i) $J = 0$ if the probability density functions are identical, $p(x|w_1) = p(x|w_2)$; (ii) $J \geq 0$; and (iii) J attains its maximum when the classes are disjoint, i.e., when $p(x|w_1) = 0$ and $p(x|w_1) \neq 0$. Some of the probabilistic distances are: average separation, Chernoff, Bhattacharyya, and the Kullback–Leibler (KL) divergence D_{KL}. This last distance is defined as [38]

$$D_{KL} = \int q(\mathbf{S}, \Psi/\mathbf{X}) \log \frac{q(\mathbf{S}, \Psi/\mathbf{X})}{p(\mathbf{S}, \Psi/\mathbf{X})} d\Psi d\mathbf{S} \tag{1.8}$$

where $\mathbf{X}$ is the set of available mixture data and $\mathbf{S}$ is the respective source vectors. Ψ denotes all the unknown parameters of the mixture data model; $p(\mathbf{S}, \Psi/\mathbf{X})$ denotes the posterior pdf for the reference model, and $q(\mathbf{S}, \Psi/\mathbf{X})$ denotes the posterior probability density function for the estimated model. We used D_{KL} as similarity measure for merging clusters in a proposed hierarchical clustering algorithm for ICA mixtures that is included in Chap. 4.

Most hierarchical clustering algorithms are variants of the so-called single-link and complete-link algorithms. These algorithms differ in the way that they characterize the similarity between a pair of clusters. In the single-link algorithm, the distance between two clusters is the minimum of the distances between all pairs of data drawn from the two clusters, whereas in the complete-link algorithm, this distance is the maximum of all pairwise distances. In either case, two clusters are merged to form a larger cluster based on minimum distance criteria. The complete-link algorithm yields more compacted clusters than those obtained by the single-link algorithm. This latter technique suffers from a chaining effect, so it has a tendency to produce clusters that are straggly or elongated [39].

As stated above, the result of the hierarchical clustering algorithm is a tree of clusters, called a dendrogram, which shows how the clusters are related. The hierarchical levels of the dendrogram can be formed in two different schemes: agglomerative (from the bottom to the upper part of the tree) and divisive (from the upper to the bottom part of the tree). An *agglomerative algorithm* begins with n subclusters, each of which contains a single data point, and then merges the two most similar groups at each stage to form a new cluster, thus reducing the number of clusters by one. The algorithm proceeds until all the data fall within a single cluster. A *divisive algorithm* starts with a single cluster composed by all the given objects and keeps splitting the clusters based on some criterion, continuing until a partition of n singleton clusters is obtained. There is a numerical value associated with each position in the tree where branches join (the distance or dissimilarity between two merged clusters).

A clustering of the data items into disjoint groups is obtained by sectioning the partition tree at a desired level. For instance, the dashed line in Fig. 1.5 at the fourth level of merging yields a clustering of three groups. This defines a data partitioning in a number of clusters of comparable homogeneity. The procedures used to find the partitioning that best fits the underlying data are called cluster validation techniques [39, 40]. In general, there are three types of criteria for evaluating of the resulting clustering structure: external criteria (a pre-specified data structure that reflects our intuition about the clustering structure); internal criteria (quantities estimated from the clusters); and relative criteria (comparing with other structures resulting from the same algorithm but with different parameter values). There are several indices proposed to implement the cluster validity techniques; they are based on measuring the compactness and separation of the clusters. Some examples of cluster validity indices are: cophenetic correlation coefficient for comparison of proximity matrices (pairwise data distances) of the resulting clustering with a known independent partition; cluster data dispersion to pairwise cluster dissimilarity measure ratio; root-mean-square standard deviation of new merged clusters; and average scattering for clusters [40]. Specifically, we have used the partition (PC) and partition entropy (PE) coefficients used from fuzzy clustering [41] to assess the partitioning obtained by the proposed hierarchical clustering algorithm of Chap. 4.

$$PC = \frac{1}{N}\sum_{i=1}^{N}\sum_{j=1}^{nc} u_{ij}^2 \tag{1.9}$$

$$PE = -\frac{1}{N}\sum_{i=1}^{N}\sum_{j=1}^{nc} u_{ij} \cdot \log_a(u_{ij}) \tag{1.10}$$

where u_{ij} denotes the degree of membership (posterior probability) of the data x_i in the j. The PC index values range from $1/nc$ to 1, where nc is the number of clusters. The closer to unity the index is, the "crisper" the clustering is. A PC value close to $1/nc$ indicates that there is no clustering tendency. The PE index values range from 0 to $\log_a(nc)$. The closer the value of PE to 0 is, the harder the clustering is. As for the PC index, the values of PE close to the upper bound $(\log_a(nc))$ indicate the absence of any clustering structure in the dataset or the inability of the algorithm to extract it.

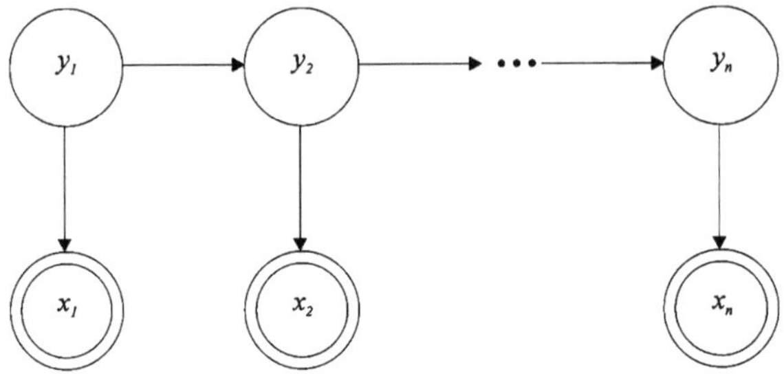

Fig. 1.6 A dynamic model (observation vector $\mathbf{x}_t$ and hidden state vector $\mathbf{y}_t$)

Hierarchical clustering algorithms have been used in a large potpourri of applications, for instance, image segmentation, object and character recognition, documental retrieval, data mining, and biomedical applications. These applications include: raster, texture, and multispectral medical image segmentation [39]; topic extraction from text corpus [42]; word sense disambiguation [43]; on-line mining of web sites usage and automatic construction of portal sites [44]; extracting financial data for business valuation [45]; grouping of non-stationary time series of industrial production indices [46]; gene expression (gene versus time, gene versus tissue, gene versus patient), interactomes (protein–protein interaction networks) and sequences (clustering protein families) [47]. A review of applications in engineering of clustering techniques can be found in [48].

1.1.4 Non-Linear Dynamic Modelling

The procedures developed in the thesis are principally focused on the analysis of ICA mixtures from static models. However, in Chap. 7 we present a procedure to extend ICAMM to the case of having sequential dependence in the feature observation record that we have called sequential ICAMM (SICAMM). We use this basis to introduce the analysis of ICA mixtures in dynamic models. Since SICAMM is defined from the classical Hidden Markov Model (HMM), the main definitions of HMM are reviewed in this section.

The basic assumption in dynamic modelling is that there exist hidden states (a hidden stochastic process) corresponding to a nonstationary model. The model can be defined using HMM having discrete states, or Kalman filters having continuous states. Figure 1.6 shows a general dynamic model with observation $\mathbf{x}_t$ and unobserved hidden state $\mathbf{y}_t$. The system is characterized by a state transition probability $P(\mathbf{y}_{t+1}|\mathbf{y}_t)$ and a state to observation probability $P(\mathbf{x}_t|\mathbf{y}_t)$.

The method for predicting future events under such a dynamic model is to maintain a posterior distribution over the hidden state $\mathbf{y}_{t+1}$ based on all observations $\mathbf{X}_{1:t} = \{\mathbf{x}_1, \ldots, \mathbf{x}_t\}$ up to time t. The posterior can be updated using the following expression

$$P(\mathbf{y}_{t+1}|\mathbf{X}_{1:t}) \propto \sum_{\mathbf{y}t} P(\mathbf{y}_t|\mathbf{X}_{1:t-1})\, P(\mathbf{x}_t|\mathbf{y}_t)\, P(\mathbf{y}_{t+1} + \mathbf{y}_t) \tag{1.11}$$

The prediction of future events $\mathbf{x}_{t+1}, \ldots, \mathbf{x}_{t+k}, k > 0$, conditioned on $\mathbf{X}_{1:t}$ is through the posterior over $\mathbf{y}_t$

$$P(\mathbf{x}_{t+1}, \ldots, \mathbf{x}_{t+k}|\mathbf{x}_{1:t}) \propto P(\mathbf{y}_{t+1}|\mathbf{X}_{1:t}) P(\mathbf{x}_{t+1}, \ldots, \mathbf{x}_{t+k}|\mathbf{y}_{t+1}) \tag{1.12}$$

Thus, the information contained in the observation $\mathbf{X}_{1:t}$ can be captured by a relatively small hidden state $\mathbf{y}_{t+1}$. Therefore, in order to predict the future, we do not have to use all previous observations $\mathbf{X}_{1:t}$ excepting its state representation $\mathbf{y}_{t+1}$. In principle, $\mathbf{y}_{t+1}$ may contain a finite history of length $k+1$, such as $\mathbf{x}_t, \mathbf{x}_{t-1} \ldots, \mathbf{x}_{t-k}$. In order to incorporate higher order dependency, a representation of the form $\mathbf{Y}_t = [\mathbf{y}'_t, \mathbf{y}'_{t-1}, \ldots, \mathbf{y}'_{t-k}]$ can be considered.

The dynamics of the system (transition and observation functions) can fit into linear or non-linear models. In the first case, the parameters of the model can be estimated using techniques such as expectation maximization (EM). Nevertheless, the system dynamics cannot be approximated linearly in many real problems. Incorporating nonlinearity to the model can be made with different methods: using probabilistic priors plus parametric models [49]; particle filtering using a finite number of samples to represent the posterior distribution that are updated with new observation arrivals [50]; and approximating the posterior $P(\mathbf{y}_t|\mathbf{X}_{1:t})$ by means of a mixture of distributions such as mixture of Gaussians (MoG) [51]. Recently, the components of the mixture have been relaxed to be non-gaussian components using ICAMM [52] in order to model the nonlinear part of the system dynamics.

HMM has evolved through multiple variations and hybrid combinations with several techniques such as neural networks, weighted transducers, and wavelets. There is an extensive range of applications for hidden state-based dynamic models. The objective is to exploit the sequential dependence of the data, which is inherent in many real-world problems, in a scheme of sequential pattern recognition. Among the HMM applications are the following: speech recognition [53, 54], video event classification and image segmentation [55], people recognition [56], human motion for robotics [57], and handwriting identification [58]. Of particular relevance is the application of HMM in event-related dynamics of brain oscillations and, in general, in causality analysis of physiological phenomena [59]. One example of these analyses is sleep staging, which is approached using radial basis function (RBF) networks and HMM for the classification of EEG recordings measured during afternoon naps in [60]. In Chap. 7, we address the sleep staging problem focusing on the analysis of arousal provoked by apnea.

1.2 Scope

The scope of this work is to explore new insights of ICA research in the context of statistical pattern recognition. Thus, new algorithms based on ICA and ICAMM are designed for feature extraction, semi-supervised learning, unsupervised hierarchical classification, sequential dynamic analysis, and modelling of particular problems. The thesis will show a more general framework for ICAMM, integrating the increase of performance and flexibility obtained by the following: non-parametric modelling of the source distributions; the incorporation of prior knowledge using semi-supervised learning; correction of non-linear dependencies in classification; use of any ICA algorithm in the learning process; construction of higher levels of classification from the ICAMM parameters; and modelling of sequential dependence in the feature observation record. The extensions are thoroughly explained and emphasized through appropriate and demonstrative examples. The results come from a large set of both synthetic and real data. The simulations and experiments are aimed at BSS and classification including comparisons of the performance and advantages of the proposed algorithms with traditional methods. Processing of the application data is off-line, and quantifiable evaluation is obtained for synthetic data where the underlying patterns and distributions are known.

In addition to the theoretical subjects, this work aims to demonstrate its suitability and usefulness with practical applications of the developed techniques in real-world problems. The research carried out in the thesis is framed within the following projects of the Signal Processing Group at Polytechnic University of Valencia:

- "Non-Linear Mixture Processor with Application in Detection, Classification, Filtering and Prediction", supported by the Spanish Administration and the FEDER Programme of the European Community under grant TEC 2008-02975.
- "Advanced algorithms for the detection-classification of ultrasonic signals", supported by the Spanish Administration and the FEDER Programme of the European Community under grant TEC 2005-01820.
- "Re-engineering of the natural stone production chain through knowledge-based processes, eco-innovation and new organisational paradigms—ISTONE", supported by The Sixth Framework Programme, European Union and the Construction Technology Institute of Valencia under grant 515762-2_IP.
- "Integration of information and communication technologies in the ambit of conservation and restoration of archaeological and ethnographic materials: non destructive testing by ultrasounds", supported by Polytechnic University of Valencia under grant PPI-05-04-5626.

The projects listed above were concerned with developing the following hardware-software prototypes: (i) a quality control system based on impact-echo testing [61] for the marble industry; and (ii) a ceramic characterization system for archaeologists. In the first prototype, the impact-echo hardware technique is empowered by suitable software for classifying the status of marble block-shaped materials building a low-cost operating system. This system might be valuable to improve the cutting of the blocks. In the second prototype, the standardization of an efficient and non-destructive method for ceramic characterization based on ultrasonic testing will be an important contribution for archaeologists. The developed prototype demonstrated that it could complement or replace destructive, costly, and time-consuming techniques, which are currently being used by archaeologists in the area of ceramic characterization. The techniques for characterization and dating of archaeological ceramics that include thermoluminescence, chemical methods, and thin section microscopy [62] are destructive, slow, and expensive.

1.3 Contributions

The thesis makes a number of contributions to the research in ICA. The fundamental contribution is a general framework for ICA mixture modelling. Further contributions are: a hierarchical method to obtain higher-level structures of classification from ICA mixture parameters; the introduction of sequential dependences in the classification of ICA mixtures; and the introduction of ICA and ICA mixtures in diverse novel applications. These applications are: material quality control using impact-echo testing, chronological cataloguing of archaeological ceramics, diagnosis of the restoration in historical buildings, diagnosis of sleep disorders, and discovery of student learning styles in web-log data.

1.3.1 ICA Mixture Modelling

A procedure is proposed that introduces the following new aspects in ICAMM: (i) Non-parametric estimation of the source pdf; (ii) Estimation of residual dependencies after ICA, and the consequent correction of the probabilities of every class to the observation vector; (iii) Supervised-unsupervised learning of the model parameters; and (iv) Incorporation of any ICA algorithm into the learning of the ICA mixture model. The final procedure was called "Mixca" (Mixture of Component Analyzers) since the independence assumption of ICA is relaxed by the proposed posterior probability correction [63].

It is assumed that every class satisfies an ICA model: vectors $\mathbf{x}_k$ corresponding to a given class C_k $k = 1, \ldots, K$ are the result of applying a linear transformation $\mathbf{A}_k$ to a (source) vector $\mathbf{s}_k$, whose elements are independent random variables, plus a bias or centroid vector $\mathbf{b}_k$, i.e., $\mathbf{x}_k = \mathbf{A}_k\mathbf{s}_k + \mathbf{b}_k$ $k = 1, \ldots, K$. Thus, Mixca is a non-linear mixture processor that fits the structure shown in Fig. 1.7. This structure has K processing channels, which are linear processors that implement the ICA equation $\mathbf{s}_k = \mathbf{A}_k^{-1}(\mathbf{x} - \mathbf{b}_k)$ followed by a non-linear processor $\mathbf{g}(\mathbf{s}_k)$.

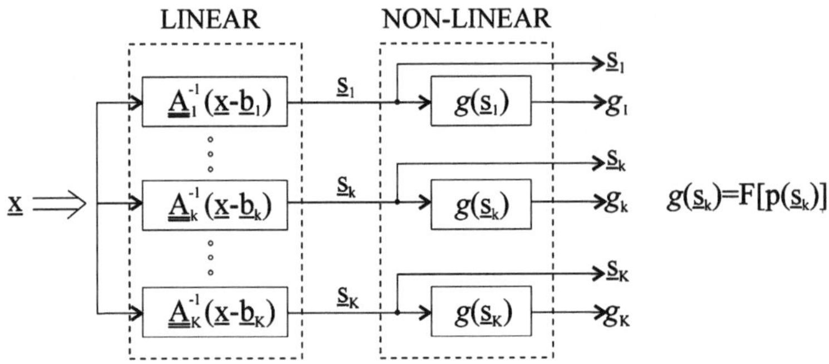

Fig. 1.7 Model for mixtures of component analyzers

Standard ICA algorithms are based on nonlinearities and higher-order statistics that try to approximate the independence property. The underlying assumptions about the source distributions and connections for some of these algorithms can be dealt with using a Maximum Likelihood framework [64]; e.g., the information maximization (InfoMax) algorithm [65] assumes that the sources are super-gaussian. In the proposed approach, the assumptions about the source distributions are relaxed as much as possible. The probability density functions of the sources are modelled with non-parametric distributions in order to increase the flexibility of the modelling and to include sources with different kinds of distributions. Recent contributions, such as Npica (non-parametric ICA) [66], which uses a kernel density estimation technique, and Radical (robust, accurate, direct ICA algorithm) [67], which is based on an entropy estimator, have demonstrated that the non-parametric ICA model is a flexible model. This model is capable of learning the source statistics and can consistently achieve an accurate separation of all the mixed signals. In the ICAMM framework, there are no references of application of non-parametric methods for source density estimation. Therefore, a non-parametric ICAMM approach is proposed with the aim of having general applicability. It does not assume any restriction on the data since the probability distributions are calculated directly from the training set through a non-parametric kernel-based approach. It also focuses the source independence directly from its definition based on the marginal distributions.

The second feature of the proposed Mixca algorithm is related to residual dependencies after ICA. There is a general consensus that if the independence hypothesis is relaxed, more general situations can be included. This can be explained by physical or heuristic arguments. Thus, statistical dependence models are considered in [68, 69], which are mainly focused on the image analysis context. In [68, 69], statistical dependencies among the variances of every component of the source vector are taken into account. In [70], it is assumed that source vectors can actually be clustered in different classes (higher level dependence) even though they have independent components. In [71], dependence is incorporated by finding clusters of tree structures across the source elements. This work takes into account possible residual dependencies among the elements of the source vectors after convergence. This accounts for possible non-linearities that are not included in the basic ICA linear model. Modelling of the residual dependencies will allow a correction of the probabilities of every class given the feature vector, thus improving the classification performance.

The third feature included in Mixca is semi-supervised learning when both labelled and unlabelled data are available to learn the model parameters. Semi-supervised learning has been extensively studied for different classification frameworks and classical generative models such as transductive support vector machines, graph-based methods, hidden Markov random fields and Bayesian classifiers [72], and in the context of the information theoretic learning [73, 74]. Recently, applications of BSS and ICA have been adapted to incorporate any available prior knowledge about the source signals (locations, statistics, and so on) into an approach called semi-blind source separation (SBSS). This is done by imposing, for instance, temporal or spatial constraints on the underlying source model [75–77]. Considering the ICAMM framework as another approach of SBSS, some observations are labelled (semi-supervised learning).

Finally, the fourth feature of Mixca is concerned with iterative solutions for ICAMM. This can be approached as conventional ICA learning for every class, with a relative degree of correction depending on the conditional probability of every class to the corresponding observation. Thus, the Mixca algorithm includes a general scheme, where any ICA algorithm can be incorporated into the learning of the model.

1.3.2 Learning Hierarchies from ICA Mixtures

A new algorithm to process the parameters (basis vectors and bias terms) learned from ICA mixtures in order to obtain hierarchical structures is proposed. The algorithm is agglomerative clustering and uses the symmetric Kullback–Leibler distance [78] to select the grouping of the clusters at each level. Meaningful higher levels of clustering can be obtained, particularly when the classes at the lowest level fit into an ICA model.

It is well known that local edge detectors can be extracted from natural scene images by standard ICA algorithms such as InfoMax [65, 79, 80] or FastIca [81–83] or new approaches such as Linear Multilayer ICA [84]. In addition, there is neurophysiological evidence that suggests a relation between the primary visual cortex activities and the detection of edges. Some theoretical dynamic models of the feedforward abstraction process from the visual cortex to higher-level abstraction have been proposed [85].

The application of the proposed method in image processing has demonstrated the algorithmic capability for merging similar patches on a natural image; clustering of different images of an object; and the creation suitable hierarchical levels of clustering from images of different objects. Furthermore the application of the

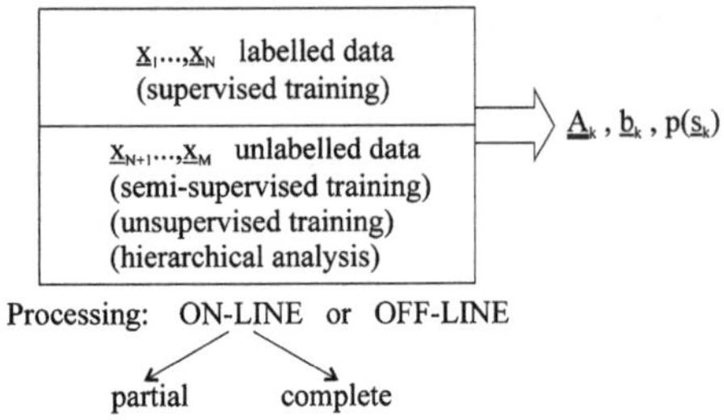

Fig. 1.8 Training and processing schemes for Mixca modelling

hierarchical algorithm in impact-echo data produced an appropriate classification tree discriminating different levels of detail for the material condition classification.

The contributions in Sects. 1.3.1 and 1.3.2 provide an extensive range of application scenarios. Figure 1.8 shows the training and processing scenarios for modelling using Mixca (Sect. 1.3.1) and hierarchical clustering defined in this section. The applications that are included in the thesis correspond to off-line processing and the different kinds of supervised training.

1.3.3 New Applications of ICA and ICAMM

The design of an accurate model to 'explain' a real physical phenomenon is a complex work. The designer has to match the assumptions and beliefs of a theoretical generative model with observed data. Actually, this in itself could be the principal objective of a thesis; however, we have not focused on this in the practical part of the thesis. On the contrary, we have pursued an exploratory objective in new fields of applications of ICA and ICAMM. Thus, we have applied the developed methods to five quite different pattern recognition problems besides the image clustering and segmentation problem explained in Sect. 1.3.2. Different levels of theoretical modelling have been approached for these problems. The novelty and usefulness of the explored applications are guaranteed since they arise from the development of research and engineering projects (see Sect. 1.2). The contributions for the real-world problems treated in the thesis are explained below.

1.3.3.1 Quality Control of Materials Using Impact-Echo Testing

In impact-echo testing, a material is hit by a hammer and the response is measured by a set of sensors located at the surface of the material [61]. The ICA-impact-echo model for materials containing point defects formulated in [86] is extended to defects with different shapes, such as cracks or holes, and the determination of the quality condition of homogeneous and defective materials is defined as an ICA mixture problem. A new model is formulated assuming the impact-echo overall scenario as a multiple-input-multiple-output linear time invariant system (MIMO-LTI) [87]. Classification of an extensive set of materials is performed at different levels of detail depending on the knowledge about the defects. It is demonstrated that the mass spectra from impact-echo testing fit ICAMM and a kind of defect signature is registered in the ICA mixture parameters. This application represents the first contribution of the application of ICAMM to NDT [88].

1.3.3.2 Chronological Cataloguing of Archaeological Ceramics

In this application, a new hardware-software system is proposed for the characterization of archaeological ceramics. The NDT methodology is based on ultrasounds. Mixca algorithm is used to classify ceramic shards using frequency and temporal features extracted from the ultrasound signal. The archaeological problem researched consisted of chronological determination of the ceramic shards. The results of classification of pieces from different deposits and ages demonstrated the best performance for Mixca even over classic methods. Mixca with a small number of labelled data (semi-supervised training) obtained higher classification accuracy than supervised methods. This is very interesting since it could be used for handling the expert uncertainty in labelling the pieces. The method has been patented and offers an alternative that could complement or replace the destructive, costly, and time-consuming techniques that are currently used.

1.3.3.3 Analysis of Restoration in Historical Buildings

The problem here consists of distinguishing between consolidated and non-consolidated zones in a restored wall of a historical building. The ultrasound signal measured is modelled with an ICA model. The mixture model considers a part of the signal from material backscattering sources and another part of the signal from sinusoidal interferences. It was possible to separate these parts and obtain improved images of the inside of the wall. The non-consolidated zone in the wall was separated from the consolidated zone using the sources extracted by ICA. In addition, the model was used for a second application consisting of the estimation of the thickness of the material layers in a historic building wall. The sources to separate were material backscattering and interferences due to instrumental noise. These applications are new in the field of ICA.

1.3.3.4 Diagnosis of Sleep Arousals in EEG Signals

The problem in this application is the detection of micro arousals that occur during sleep at night due to apnea. Modelling of dynamic changes in the ICA mixtures by including temporal dependencies in classification is proposed (that we called

SICAMM). The model consists of a sequential Bayes processor formulated from HMM theory. The parameters estimated by Mixca (mixture matrices, centroids, and source probability densities) in conjunction with the class-transition probabilities are used for classification. SICAMM is applied in simulations and real data analysis. This analyses consists of processing features extracted from 8-hours EEG signals to estimate a two-class (wake and sleep) hypnogram [89]. Both simulated and real data demonstrate the potential interest of including sequential dependence in the implementation of an ICAMM classifier. Thus, a more accurate detection of arousals in the hypnogram will help to the medical diagnosis of sleep disorders.

1.3.3.5 Discovering of Learning Styles in E-Learning

This application introduces the use of ICA as a data mining technique to extract patterns of academic learning styles from a virtual campus. The analysis is based on a well-known academic model (Felder's model) that classifies the learning style of students according to the ways in which they receive and process information [90]. Mixca is configured to estimate one ICA. It is demonstrated that the mixing matrix could be used to associate dimensions of learning and web learning activities from a huge amount (more than 2.3 millions of records) of historical web log data. The learning styles detected automatically were consistent with the courses and teaching methodologies of the e-learning campus. The results obtained by ICA integrated with the results obtained by clustering and decision trees allow an academic action outline for improving the learning process to be designed.

1.4 Overview

The thesis is presented in eight chapters. Excluding the Introduction (this chapter) and Conclusions and Future Directions (Chap. 8), the thesis can be divided into three parts: background (Chap. 2), theoretical contributions (Chaps. 3 and 4), and new applications (Chaps. 5, 6 and 7). Chapter 2 deals with theoretical foundations about ICA and ICA mixtures. The performance of the proposed methods is compared with selected algorithms that are representative of the many different types of ICA and ICAMM algorithms that exist in the literature [13–18]. In order to assist with these comparisons, a review of the selected algorithms is included.

The central contribution that this thesis makes is in Chap. 3, which formulates a general framework for modelling mixtures of ICAs. This chapter introduces the ICA mixture model where each observation vector corresponds to a class which is defined by an ICA model. Then the problem is stated as how to estimate the mixture matrix and the bias terms for each class. This problem is approached by a log-likelihood cost function of the unknowns, and an optimization procedure using natural gradient algorithm is formulated. Four new features are included in the resulting method for learning the ICA mixtures: non-parametric estimation of the source pdf's, semi-supervised learning, use of any ICA algorithm in the parameter updating, and a correction of the posterior probability after convergence. Afterward, this chapter explores the performance of the algorithm by an extensive set of simulations and experiments. The simulations include: performance in BSS, classification of ICA mixtures, classification of ICA mixtures with nonlinear dependencies, and semi-supervised learning. The results are discussed and compared with several standard ICA algorithms.

Chapter 4 describes a postprocessing method to be applied after obtaining the parameters of the ICA mixtures by a non-parametric method as the one described in Chap. 3. The method consists of a hierarchy algorithm that composes an agglomerative (bottom-up) clustering from the estimated parameters (basis vectors and bias terms) of the ICA mixture. The merging at different levels of the hierarchy is performed using the Kullback–Leibler distance between clusters. The method is validated from several simulations (including ICA mixtures with uniform and Laplacian source distributions) and from processing real data. The applications are image processing segmentation and clustering, and classification of materials tested using impact-echo. Meaningful hierarchical levels are demonstrated for the experiments from the ICA mixture parameters to higher-level structures. The hierarchical levels for image processing represent concepts of similarity in a set of images of different objects or in patches of an image.

Novel applications of the algorithmic methods developed in Chaps. 3 and 4 are covered in the following chapters. Chapters 5 and 6 are devoted to NDT applications, and Chap. 7 is dedicated to biosignal processing and webmining. Chapter 5 presents a model for the classification of materials evaluated by the impact-echo testing technique. It is demonstrated that several kinds of defects are characterized by the parameters of different ICA models from the mass spectra of the impact-echo test. This modelling, allows the following levels of classifications: (i) *Material condition* homogeneous, one defect, multiple defects; (ii) *Kind of defect* homogeneous, hole, crack, multiple defects; (iii) *Defect orientation* homogeneous, hole in the X axis or Y axis, crack in the XY, ZY, or XZ planes, multiple defects; and (iv) *Defect dimension* homogeneous, passing through or half-passing through holes and cracks of classification level (iii), multiple defects. The chapter includes results from a large number of lab experiments. The performance of the classification by ICA mixtures using Mixca is compared with linear discriminant analysis (LDA) and with multi-layer perceptron (MLP) classification.

Chapter 6 includes two applications in the NDT field. The first application comprises the classification of several archaeological ceramic shards from different deposits in the eastern part of Spain. The pieces are measured by ultrasounds using an *ad hoc* device. The features extracted from the ultrasonic signals are classified using the following algorithms: LDA, MLP, RBF, learning vector quantization (LVQ), k-nearest neighbours (kNN), and Mixca. The classification performed was *chronological period*: the Bronze Age, Iberian, Roman, and the Middle Ages. The best performance in classification is obtained for different ratios of semi-supervised training of Mixca. Some of the pieces were characterized using physical and chemical analyses. A rationale of the results that shows the relationship between ceramic physical properties, ultrasound propagation, and method employed to manufacture the pieces is included.

In the second application of Chap. 6, the ICA mixture algorithm is applied as a non-parametric one-ICA algorithm for BSS. The goals are to evaluate the restoration consolidation and to detect interfaces in a wall of a historical dome of a Basilica using ultrasounds. The measured signals contain the contribution of the injected ultrasonic pulse buried in backscattering grain noise plus the contribution of sinusoidal phenomena. The sources and mixture matrix extracted by ICA allow these contributions to be separated. The recovered sinusoidal sources characterize the resonance phenomenon of multiple reflections of the ultrasonic pulse at non-consolidated zones and instrument interferences.

Chapter 7 presents two different applications. First, a new model for sequential pattern recognition based on HMM and ICAMM is proposed. The model is called SICAMM. It is applied to the first application of Chap. 7, which is the detection of micro arousals caused by apnea during the night. These abrupt changes are registered in a diagram called hypnogram, which shows the transitions between the sleep stages. Long EEG records from apnea patients measured during sleep are analyzed. Two sleep stages are classified: wake and sleep. The classification obtained by SICAMM outperforms the ICAMM classification showing accuracy for detection of episodes of wakefulness during sleep. The second application of Chap. 7 is webmining from a huge amount of historical web log data from e-learning activities. The ICA mixture algorithm is configured to estimate the parameters for only one ICA. The application consists of the detection of student learning styles based on a known educational framework. The mixture matrix obtained by ICA demonstrates the relation between e-learning style dimensions and e-learning web activities leading to the detection of student learning styles.

Finally, Chap. 8 ends the thesis with the conclusions and findings, discussion of the open relevant subjects, and discussion about the future directions of research.

References

1. C.J.D.M. Verhagen, Some general remarks about pattern recognition: its definition; its relation with other disciplines; a literature survey. Pattern Recognit. **7**(3), 109–116 (1975)
2. J.C. Bezdek, S.K. Pal, *Fuzzy Models for Pattern Recognition: Methods That Search for Structures in Data* (IEEE press, New York, 1992)
3. H.A. Simon, Science Seeks Parsimony, Not Simplicity: Searching for Pattern In Phenomena. in *Simplicity, Inference and Modelling* (*Keeping it Sophisticatedly Simple*), ed. by A. Zellner, H.A. Keuzenkamp, M. McAleer (Cambridge University Press, Cambridge, 2004)
4. V. Castelli, T.M. Cover, The relative value of labelled and unlabelled samples in pattern recognition with an unknown mixing parameter. IEEE Trans. Inf. Theory **42**(6), 2102–2117 (1996)
5. I. Cohen, F.G. Cozman, N. Sebe, M.C. Cirelo, T.S. Huang, Semisupervised learning of classifiers: theory, algorithms, and their application to human-computer interaction. IEEE Trans. Pattern Anal. Mach. Learn. **26**(12), 1553–1567 (2004)
6. O. Chapelle, B. Schölkopf, *A Semi-supervised Learning* (MIT Press, Zien, 2006)

7. S. Haykin, *Intelligent Signal Processing* (Wiley-IEEE Press, New York, 2001)
8. A.K. Jain, Statistical pattern recognition: a review. IEEE Trans. Pattern Anal. Mach. Intell. **22**(1), 4–37 (2000)
9. A.R. Webb, *Statistical Pattern Recognition* (Wiley, New York, 2002)
10. T.W. Lee, M.S. Lewicki, T.J. Sejnowski, ICA mixture models for unsupervised classification of non-gaussian classes and automatic context switching in blind signal separation. IEEE Trans. Pattern Anal. Mach. Intell. **22**(10), 1078–1089 (2000)
11. R. Choudrey, S. Roberts, Variational mixture of bayesian independent component analysers. Neural Comput. **15**(1), 213–252 (2002)
12. P. Comon, Independent component analysis—a new concept? Signal Process. **36**(3), 287–314 (1994)
13. A. Hyvärinen, J. Karhunen, E. Oja, *Independent Component Analysis* (Wiley, New York, 2001)
14. A. Cichocki, S. Amari, *Adaptive Blind Signal and Image Processing: Learning Algorithms and Applications* (Wiley, New York, 2001)
15. T.W. Lee, *Independent Component Analysis—Theory and Applications* (Kluwer Academic Publishers, Boston, 1998)
16. S. Roberts, R. Everson, *Independent Component Analysis—Principles and Practice* (Cambridge University Press, Cambridge, 2001)
17. A. Cichocki, R. Zdunek, A.H. Phan, S. Amari, *Nonnegative Matrix and Tensor Factorizations: Applications to Exploratory Multi-way Data Analysis and Blind Source Separation* (Wiley, New York, 2009)
18. P. Comon, C. Jutten (eds.), *Handbook of Blind Source Separation Independent Component Analysis and Applications* (Academic Press, Oxford, 2010)
19. V. Zarzoso, A. Nandi, Noninvasive fetal electrocardiogram extraction: blind separation versus adaptive noise cancellation. IEEE Trans. Biomed. Eng. **48**(1), 12–18 (2001)
20. J.J. Rieta, F. Castells, C. Sanchez, V. Zarzoso, Atrial activity extraction for atrial analysis using blind source separation. IEEE Trans. Biomed. Eng. **51**(7), 1176–1186 (2004)
21. R. Llinares, J. Igual, A. Salazar, A. Camacho, Semi-blind source extraction of atrial activity by combining statistical and spectral features. Digit. Signal Process. **21**(2), 391–403 (2011)
22. S. Makeig, J. Onton, in *A Trial-by-Trial Pattern Approach to Event-Related EEG Analysis: ERP Features and EEG Dynamics: An ICA Perspective*, ed. by S. Luck, E. Kappenman. Oxford Handbook of Event-Related Potential Components (Oxford University Press, Oxford, 2009)
23. D.M. Groppe, S. Makeig, M. Kutas, Identifying reliable independent components via split-half comparisons. Neuroimage **45**, 1199–1211 (2009)
24. Y. Karklin, M.S. Lewicki, Learning higher-order structures in natural images. Netw. Comput. Neural Syst. **14**, 483–499 (2003)
25. T.W. Lee, M.S. Lewicki, Unsupervised image classification, segmentation, and enhancement using ICA mixture models. IEEE Trans. Image Process. **11**(3), 270–279 (2002)
26. N.H. Mollah, M. Minami, S. Eguchi, Exploring latent structure of mixture ICA models by the Minimum ß-Divergence method. Neural Comput. **18**, 166–190 (2005)
27. C.A. Shah, M.K. Arora, P.K. Varshney, Unsupervised classification of hyperspectral data: an ICA mixture model based approach. Int. J. Remote Sens. **25**(2), 481–487 (2004)
28. C.A. Shah, P.K. Varshney, M.K. Arora, ICA mixture model algorithm for unsupervised classification of remote sensing imagery. Int. J. Remote Sens. **28**(8), 1711–1731 (2007)
29. U.M. Bae, T.W. Lee, S.Y. Lee, Blind signal separation in teleconferencing using the ICA mixture model. Electron. Lett. **37**(7), 680–682 (2000)
30. K. Chan, T.W. Lee, T.J. Sejnowski, Variational learning of clusters of undercomplete nonsymmetric independent components. J. Mach. Learn. Res. **3**, 99–114 (2002)
31. M.H. Goldbaum, A.P.A. Sample, Z. Zhang, K. Chan, J. Hao, T.W. Lee, C. Boden, C. Bowd, R. Bourne, L. Zangwill, T. Sejnowski, D. Spinak, R.N. Weinreb, Using unsupervised learning with independent component analysis to identify patterns of glaucomatous visual field defects. Investig. Ophthalmol. Vis. Sci. **46**(10), 3676–3683 (2005)

32. T.P. Jung, S. Makeig, T.W. Lee, M. J. McKeown, G. Brown, A. J. Bell, T. J. Sejnowski, Independent component analysis of biomedical signals. *Proceedings of the 2nd International Workshop on Independent Component Analysis and Signal Separation*, pp. 633–644, 2000
33. C.T. Lin, W.C. Cheng, S.F. Liang, An on-line ICA-mixture-model-based self-constructing fuzzy neural network. IEEE Trans. Circuits Syst. **52**(1), 207–221 (2005)
34. X. Hu, A. Shimizu, H. Kobatake, S. Nawano, Aplying ICA mixture analysis for segmenting liver from multi-phase abdominal CT images. Lect. Notes Comput. Sci. **3150**, 54–61 (2004)
35. I. Guyon, A. Elisseeff, An introduction to variable and feature selection. J. Mach. Learn. Res. **3**, 1157–1182 (2003)
36. L. Devroye, L. Györfi, G. Lugosi, *A Probabilistic Theory of Pattern Recognition* (Springer, New York, 1996)
37. B.W. Silverman, *Density Estimation for Statistics and Data Analysis* (Chapman and Hall, London, 1985)
38. D.J. Mackay, *Information Theory, Inference, and Learning Algorithms* (Cambridge University Press, Cambridge, 2004)
39. A.K. Jain, M.N. Murty, P.J. Flynn, Data clustering: a review. ACM Comput. Surv. **31**(3) (1999)
40. M. Haldiki, Y. Batistakis, M. Vazirgiannis, On clustering validation techniques. J. Intell. Inf. Syst. **17**(2–3), 107–145 (2001)
41. J.C. Bezdek, *Pattern Recognition with Fuzzy Objective Function Algorithms* (Plenum Press, New York, 1981)
42. J. Zeng, C. Wu, W. Wang, Multi-grain hierarchical topic extraction algorithm for text mining. Expert Syst. Appl. **37**, 3202–3208 (2010)
43. R. Navigli, Word sense disambiguation: a survey. ACM Comput. Surv. **41**(2), 10:1–10:69 (2009)
44. H. Azzag, G. Venturini, A. Oliver, C. Guinot, A hierarchical ant based clustering algorithm and its use in three real-world applications. Eur. J. Oper. Res. **179**, 906–922 (2007)
45. J.L. Seng, J.T. Lai, An Intelligent information segmentation approach to extract financial data for business valuation. Expert Syst. Appl. **37**, 6515–6530 (2010)
46. J.A. Vilar, A.M. Alonso, J.M. Vilar, Non-linear time series clustering based on non-parametric forecast densities. Comput. Stat. Data Anal. **54**, 2850–2865 (2010)
47. B. Andreopoulos, A. An, X. Wang, M. Schroeder, A roadmap of clustering algorithms: finding a match for a biomedical application. Briefings Bioinf. **10**(3), 297–314 (2009)
48. D.T. Pham, A.A. Afify, Clustering techniques and their applications in engineering. Proc. Inst. Mech. Eng. Part C J. Mech. Eng. Sci. **221**(11), 1445–1459 (2007)
49. S. Roweis, Z. Ghahramani, in *Learning Nonlinear Dynamical Systems Using the EM Algorithm*, ed. by S. Haykin. Kalman Filtering and Neural Networks (Wiley, New York, 2001), pp. 175–220
50. J.V. Candy, Bootstrap particle filtering. IEEE Signal Process. Mag. **24**(4), 73–85 (2007)
51. R. Chen, J.S. Liu, Mixture Kalman filters. J. Royal Stat. Soc. Ser. B **62**, 493–508 (2000)
52. J. Zhou, X.P. Zhan, Hidden Markov Model Framework Using Independent Component Analysis Mixture Model. *Proceedings of the 31st IEEE International Conference on Acoustics, Speech, and Signal Processing*, pp. V:553–556, Toulouse, 2006
53. L.R. Rabiner, A tutorial on hidden Markov models and selected applications in speech recognition. Proc. IEEE **77**(2), 257–286 (1989)
54. D.H. Milone, L.E. Di Persia, M.E. Torres, Denoising and recognition using hidden Markov models with observation distributions modeled by hidden Markov trees. Pattern Recognit. **43**, 1577–1589 (2010)
55. X. Ma, D. Schonfeld, A.A. Khokhar, Video event classification and image segmentation based on noncausal multidimensional hidden Markov Models. IEEE Trans. Image Process. **18**(6), 1304–1313 (2009)
56. F. Matta, J.L. Dugelay, Person recognition using facial video information: a state of the art. J. Vis. Lang. Comput. **20**, 180–187 (2009)

57. D. Kulic, W. Takano, Y. Nakamura, Online segmentation and clustering from continuous observation of whole body motions. IEEE Trans. Robot. **25**(5), 1158–1166 (2009)
58. Z. He, X. You, Y.Y. Tang, Writer identification of Chinese handwriting documents using hidden Markov tree model. Pattern Recognit. **41**(4), 1295–1307 (2008)
59. C. Neuper, W. Klimesch, *Event-Related Dynamics of Brain Oscillations: Progress in Brain Research Series*, vol 159 (Elsevier, Amsterdam, 2006)
60. J. Kohlmorgen, K.R. Muëller, J. Rittweger, K. Pawelzik, Identification of nonstationary dynamics in physiological recordings. Biol. Cybern. **83**(1), 73–84 (2000)
61. M. Sansalone, W.B. Streett, *Impact-Echo: Non-Destructive Evaluation of Concrete and Masonry* (Bullbrier Press, Ithaca, 1997)
62. R.E. Taylor, M.J. Aitken, *Chronometric Dating in Archaeology: Advances in Archaeological and Museum Science Series*, vol. 2 (Springer, New York, 1997)
63. A. Salazar, L. Vergara, A. Serrano, J. Igual, A general procedure for learning mixtures of independent component analyzers. Pattern Recognit. **43**(1), 69–85 (2010)
64. J.F. Cardoso, High-order contrasts for independent component analysis. Neural Comput. **11**(1), 157–192 (1999)
65. A.J. Bell, T.J. Sejnowski, An information-maximization approach to blind separation and blind deconvolution. Neural Comput. **7**, 1129–1159 (1995)
66. R. Boscolo, H. Pan, Independent component analysis based on nonparametric density estimation. IEEE Trans. Neural Netw. **15**(1), 55–65 (2004)
67. E.G. Learned-Miller, J.W. Fisher, ICA using spacings estimates of entropy. J. Machine Learn. Res. **4**, 1271–1295 (2003)
68. A. Hyvärinen, P.O. Hoyer, M. Inki, Topographic independent component analysis. Neural Comput. **13**(7), 1527–1558 (2001)
69. A. Hyvärinen, J. Hurri, Blind separation of sources that have spatiotemporal variance dependencies. Signal Process. Special Issue Indep. Compon. Anal. Beyond **84**(2), 247–254 (2004)
70. H.J. Park, T.W. Lee, Capturing nonlinear dependencies in natural images using ICA and mixture of Laplacian distribution. Neurocomputing **69**, 1513–1528 (2006)
71. F.R. Bach, M.I. Jordan, Beyond independent components: trees and clusters. J. Mach. Learn. Res. **3**, 1205–1233 (2003)
72. O. Chapelle, B. Schölkopf, A. Zien, *Semi-supervised Learning* (MIT Press, Cambridge, 2006)
73. K.H. Jeong, J.W. Xu, D. Erdogmus, J.C. Principe, A new classifier based on information theoretic learning with unlabelled data. Neural Netw. **18**, 719–726 (2005)
74. D. Erdogmus, J.C. Principe, From linear adaptive filtering to nonlinear information processing—the design and analysis of information processing systems. IEEE Signal Process. Mag. **23**(6), 14–33 (2006)
75. C.W. Hesse, C.J. James, On semi-blind source separation using spatial constraints with applications in EEG Analysis. IEEE Trans. Biomed. Eng. **53**(12–1), 2525–2534 (2006)
76. J. Even, K. Sugimoto, An ICA approach to semi-blind identification of strictly proper systems based on interactor polynomial matrix. Int. J. Robust Nonlinear Control **17**, 752–768 (2007)
77. Z. Ding, T. Ratnarajah, C.F.N. Cowan, HOS-based semi-blind spatial equalization for MIMO rayleigh fading channels. IEEE Trans. Signal Process. **56**(1), 248–255 (2008)
78. L. Vergara, J. Gosálbez, J.V. Fuente, R. Miralles, I. Bosch, A. Salazar, A. Lopez, L. Domínguez, Ultrasonic nondestructive testing on marble block rocks. Mater. Evaluation **62**(1), 73–78 (2004)
79. T.W. Lee, M. Girolami, T.J. Sejnowski, Independent component analysis using an extended InfoMax algorithm for mixed sub-gaussian and super-gaussian sources. Neural Comput. **11**(2), 417–441 (1999)
80. A.J. Bell, T.J. Sejnowski, The "Independent Components" of natural scenes are edge filters. Vis. Res. **37**(23), 3327–3338 (1997)
81. A. Hyvärinen, E. Oja, A fast fixed-point algorithm for independent component analysis. Neural Comput. **9**(7), 1483–1492 (1998)

82. A. Hyvarinen, Fast and robust fixed-point algorithms for independent component analysis. IEEE Trans. Neural Netw. **10**(3), 626–634 (1999)
83. J.H. Van Hateren, A. van der Shaaf, Independent component filters of natural images compared with simple cells in primary visual cortex. Proc. R. Soc. Lond. B **265**, 359–366 (1998)
84. Y. Matsuda, K. Yamaguchi, Linear multilayer ICA generating hierarchical edge detectors. Neural Comput. **19**(1), 218–230 (2007)
85. T.S. Lee, D. Mumford, Hierarchical Bayesian inference in the visual cortex. J. Opt. Soc. Am. A **20**(7), 1434–1448 (2003)
86. A. Salazar, L. Vergara, J. Igual, J. Gosalbez, Blind source separation for classification and detection of flaws in impact-echo testing. Mech. Syst. Signal Process. **19**(6), 1312–1325 (2005)
87. Y. Huang, J. Benesty, J. Chen, *Acoustic MIMO Signal Processing* (Springer, Berlin, 2006)
88. A. Salazar, L. Vergara, R. Llinares, Learning material defect patterns by separating mixtures of independent component analyzers from NDT Sonic Signals. Mech. Syst. Signal Process. **24**(6), 1870–1886 (2010)
89. M. Jobert, H. Shulz, P. Jähnig, C. Tismer, F. Bes, H. Escola, A computerized method for detecting episodes of wakefulness during sleep based on the Alpha slow-wave index (ASI). Sleep **17**(1), 37–46 (1994)
90. R. Felder, L. Silverman, Learning and teaching styles. J. Eng. Educ. **78**(7), 674–681 (1988)

Chapter 2
ICA and ICAMM Methods

2.1 Introduction

The seminal work of the research in ICA was provided by Jutten in [1–4]. Independent component analysis (ICA) aims to separate hidden sources from their observed linear mixtures without any prior knowledge. The only assumption about the sources is that they are mutually independent [5]. Thus, the goal is blind source estimation; although it has been recently alleviated by incorporating prior knowledge about the sources into the ICA model in the so-called semi-blind source separation (see for instance [6–8]). This technique has been widely used in many fields of application such as telecommunications, bioengineering, and material testing [5]. There is extensive literature that reviews and provides taxonomies and comparisons about the large number of ICA algorithms that have been developed during the last two decades (see for example [5, 9–13]). Therefore, in this chapter, instead of undertaking an exhaustive review of the methods, we will focus on reviewing the following: the ICA basic concepts, some ICA algorithms that will be used for comparison with those proposed in this work, and existing ICAMM algorithms.

The standard noiseless instantaneous ICA formulates a Mx1 random vector $\mathbf{x}$ by linear mixtures of M random variables that are mutually independent $s_1, \ldots, s_M$ whose distributions are totally unknown. That is, for $\mathbf{s} = (s_1, \ldots, s_M)^T$ and some matrix $\mathbf{A}$

$$\mathbf{x} = \mathbf{A}\mathbf{s} \tag{2.1}$$

The essential principle is to estimate the so-called mixing matrix $\mathbf{A}$, or equivalently $\mathbf{B} = \mathbf{A}^{-1}$ (the demixing matrix). The matrix $\mathbf{A}$ contains the coefficients of the linear transformation that represents the transfer function from sources to observations. Thus, given N i.i.d. observations $(\mathbf{x}_1, \ldots, \mathbf{x}_N)$ from the distribution of $\mathbf{x}$, A^{-1} can be applied to separate each of the sources $\mathbf{s}_i = \mathbf{B}_i\,\mathbf{x}$, where $\mathbf{B}_i$ is the ith

A. Salazar, *On Statistical Pattern Recognition in Independent Component Analysis Mixture Modelling*, Springer Theses, DOI: 10.1007/978-3-642-30752-2_2,

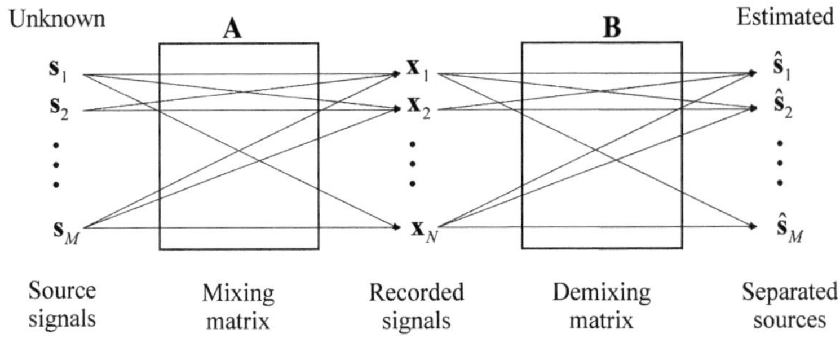

Fig. 2.1 The instantaneous mixing and unmixing model for BSS-ICA

row of **B**. This can be seen as a projection pursuit density estimation problem to find M directions such that the corresponding projections are the most mutually independent. For the sake of simplicity, we will assume the square problem (the same number of sources as mixtures, thus the order of **A** is MxM). Figure 2.1 shows a schema that illustrates the instantaneous mixing and unmixing model for BSS-ICA.

Furthermore, the instantaneous linear model can be applied in the frequency domain for the analysis of convolutive mixtures. Applying the Fourier transform to both sides of Eq. (2.1), we obtain the following frequency expression

$$\mathbf{x}(\omega) = \mathbf{A}(\omega)\mathbf{s}(\omega) \tag{2.2}$$

where $\mathbf{x}(\omega) = \mathrm{FT}\{\mathbf{x}\}$, $\mathbf{A}(\omega) = \mathrm{FT}\{\mathbf{A}\}$, and $\mathbf{s}(\omega) = \mathrm{FT}\{\mathbf{s}\}$ are the Fourier transforms of the observation vector, mixing matrix, and source vector, respectively. The time and frequency domain ICA models are equivalent, but the coefficients of the transfer matrix may vary with ω (see for instance [14] and the references within). An attempt to generalize the BSS algorithms for MIMO signal processing that exploits three signal properties nonwhitenes, nongaussianity, and nonstationarity in an information theoretic cost function has been recently formulated in [15, 16]. In some cases, the convolutive model can be solved as an ''instantaneous'' problem for selected frequencies. The frequency component permutation problem is thus avoided. The frequencies to be analyzed are selected according to the application; for instance, in a detection problem, the frequencies around the working frequency of the excitation sensor are in the band of interest. We include an example of this frequency ICA analysis applied in NDT in Chap. 5.

It is well-known that **A** is identifiable, up to scaling and permutation of columns, when **s** has at most one Gaussian component and **A**is assumed to be nonsingular [17] The restriction in Gaussian components is explained by the central limit theorem, considering that a linear mixture of independent random variables is more Gaussian than the original variables. Thus, to specify **B** uniquely, we need to put some scale and permutation constraints either on **s** or on **B**. Because of the ICA indeterminacies the sources are usually assumed to be unit variance. Also, it is commonly assumed that both the observed variables and the independent components have zero mean.

The source independence is expressed as the joint probability, which is the product of the marginal densities $p(\mathbf{s}) = \prod_i p_i(\mathbf{s}_i)$. Since the source distribution is not available, the independence is represented in different ways, e.g., using the following statistics

$$E\left[g_i(\hat{\mathbf{s}}_\mathbf{i})g_j\left(\hat{\mathbf{s}}_j\right)\right] = 0 \tag{2.3}$$

for any non-linear function g_i, i.e., all the cross cumulants must be zero.

Most of the existing algorithms used to estimate the matrix **A** can be organized in two categories. The first category of methods directly approximates the distributions of hidden sources within a specified class of distributions and minimizes a cost function the so-called contrast function, or simply contrast, which is generically denoted $\phi(\hat{\mathbf{s}})$ such as mutual information, likelihood function, or equivalents [5, 17–21]. The design of the ICA algorithms includes the formulation of a contrast function that has to be minimized through an optimization procedure. The contrast function is a real valued function of the estimated sources **s**, which yields a minimum value when the independence is attained. The second category of methods optimizes other contrast functions without approximating distributions explicitly. These functions can be, for instance, nongaussianity (using neguentropy or kurtosis), and nonlinear correlation among estimated sources [2, 22].

In several ICA algorithms, the data are first whitened (also called sphering), which requires the covariance matrix of the data to be unity. It is well-known that the demixing matrix can be factorized as the product of a whitening and an orthogonal matrix, i.e., $\mathbf{B} = \mathbf{VW}$, where **V** is the whitening matrix and **W** is the orthogonal one. The mixtures are first whitened in order to exhaust the second order moments (signals are forced to be uncorrelated). The whitened vector is expressed as $\mathbf{z} = \mathbf{VAs}$, with $E = \left[\mathbf{zz}^T\right] = \mathbf{I}$, and the whiteness constraint $E\left[\hat{\mathbf{s}}\hat{\mathbf{s}}^T\right] = \mathbf{I}$., with $\hat{\mathbf{s}}$ being the estimated sources. Thus, the ICA model, considering a prewhitening step, is expressed as

$$\hat{\mathbf{s}} = \mathbf{Bx} = \mathbf{WVx} \tag{2.4}$$

The orthogonal matrix **W** is a rotation of the joint density, which has to maximize the nongaussianity of the marginal densities, thus maximizing a measure of independence. The rotation step keeps the covariance of $\hat{\mathbf{s}}$ equal to the identity, thus preserving the whiteness, hence, the decorrelation of the components. Prewhitening is an optional step to estimate the ICA parameters; in fact, recent methods avoid a prewhitening phase and directly attempt to compute a non-orthogonal diagonalizing congruence (see e.g., [23, 24]. A discussion about connections between mutual information, entropy, and non Gaussianity in a general framework without imposing whitening is presented in [25]. However, prewhitening in ICA algorithms has been reported to provide algorithmic computational advantages (see e.g., [26, 27]).

The algorithms used in ICA can be deterministic or stochastic. The deterministic algorithms always produce the same results (usually exploiting the algebraic structure of the matrices involved) whereas the stochastic algorithms are adaptive starting from a random unmixing matrix that is updated iteratively. The updating can be made for every observation (on-line) or for the whole set of observations (off-line). Thus, the results of stochastic algorithms vary in different executions of the algorithm. The reliability of the results has to be studied since the algorithm may reach a local optimum (local consistency) instead of the unique global optimum (global consistency) of the contrast function. The convergence depends on statistical variables such as random sampling of the data. It is commonly accepted that the estimation results are robust to the details of knowledge about the distributions (super- or sub-gaussianity, and so on). It has also been demonstrated that incorrect assumptions on such distributions can result in poor estimation performance, and sometimes in a complete failure to obtain the source separation [28]. Local consistency of ICA methods that search for specified distributions and global consistency in the case of two sources with heavy-tail distributions has been studied [19, 26, 29]. Recently, the statistical reliability or "quality" of the parameters estimated by ICA has been analyzed using bootstrap resampling techniques and visualization of the cluster structure of the components [30, 31].

2.2 Standard ICA Methods

The ideal measure of independence is the "mutual information" that was proposed as a contrast function in [17]. It has been demonstrated that this function corresponds to the likelihood for a model of independent components that is optimized with respect to all its parameters. Thus, the likelihood in a given ICA model is the probability of a data set as a function of the mixing matrix and the component distributions [28]. Mutual information (I) is defined as the Kullback–Leibler (KL) divergence or relative entropy between the joint density and the product of the marginal distributions:

$$I(\hat{\mathbf{s}}) = KL\left(\hat{\mathbf{s}}; \prod_i p(\hat{s}_i)\right) = \int p(\hat{\mathbf{s}}) \log \frac{p(\hat{\mathbf{s}})}{\prod_i p(\hat{s}_i)} d\hat{\mathbf{s}} \tag{2.5}$$

It is non-negative and equals to zero only if the distributions are the same. The logarithm of the fraction in Eq. (2.5) can be transformed into a difference of logarithms, obtaining

$$I(\hat{\mathbf{s}}) = \sum_i H(\hat{s}_i) - H(\hat{\mathbf{s}}) \tag{2.6}$$

where $H(u)$ denotes Shannon's differential entropy for a continuous random variable u, which can be seen as a measure of the randomness of the variable u.

$$H(u) = -\int p(u)\log p(u)du \tag{2.7}$$

The entropy of the estimated sources $H(\hat{\mathbf{s}})$ in Eq. (2.4) equals $H(\mathbf{x}) - \log|\det \mathbf{B}|$. If a step of prewhitening is considered, all the white versions of $\hat{\mathbf{s}}$ are rotated versions of each other and $\log|\det \mathbf{B}| = 0$ since $\mathbf{B}$ is an orthogonal matrix. For this case, the entropy $H(\hat{\mathbf{s}})$ remains constant and, thus, the mutual information (or dependence) is equal to the sum of the marginal entropies of $\hat{\mathbf{s}}$ (up to the constant term $H(\hat{\mathbf{s}})$)

$$I(\mathbf{s}) = \sum_i H(\hat{s}_i) \Leftrightarrow E\left[\hat{\mathbf{s}}\hat{\mathbf{s}}^T\right] = \mathbf{I} \tag{2.8}$$

Thus, there is a connection between maximum independence and minimum entropy; the objective of maximizing the independence is equivalent to the objective of minimizing the sum of the entropies of all components. In this sense, ICA is a minimum entropy method under the whitening constraint $E\left[\hat{\mathbf{s}}\hat{\mathbf{s}}^T\right] = \mathbf{I}$. In addition, the entropy $-H(u)$ is equal to the Kullback–Leibler divergence between the random variable u and the zero mean unit variance Gaussian density (up to a constant). Hence, the mutual information contrast imposes finding marginal distributions as far as possible from Gaussianity. Furthermore, it has been demonstrated that the mutual information can be decomposed under linear transforms as the sum of two contributions: a contribution expressing the decorrelation of the components and a contribution expressing their non Gaussianity [25].

Unfortunately, the mutual information is difficult to approximate and optimize on the basis of a finite sample; thus, much research on ICA has focused on alternative solutions [17, 18, 20]. A popular approach for estimating the ICA model is the maximum likelihood (ML) estimation. The goal is to find the ICA parameters that give the highest probability for the observed data. The problem is formulated as $p_x(\mathbf{x}) = |\det \mathbf{B}|p_s(\mathbf{s}) = |\det \mathbf{B}|\prod_i p_i(s_i)$, where p_x is the density of the mixture vector, and p_i denotes the density of the independent components. Considering N samples available to evaluate the densities, the maximum log likelihood for the ICA model can be defined as

$$\frac{1}{N}\log L(\mathbf{B}) = E\left[\sum_i \log\ p_i\left(b_i^T\mathbf{x}\right)\right] + \log|\det \mathbf{B}| \tag{2.9}$$

where the expectation $E[\cdot]$ is the average computed from the observed samples.

Several methods to define the contrast function have been proposed in the literature. These methods are based on non Gaussianity, mutual information, higher order statistics (cumulants), and time structures [5]. The contrasts are closely connected, and have been implemented in different ICA algorithms for BSS with successful applications in many fields. The optimization techniques applied to the contrast function for adaptive algorithms are mainly based on gradient (natural, descent, etc.) and approximate Newton methods. The estimation

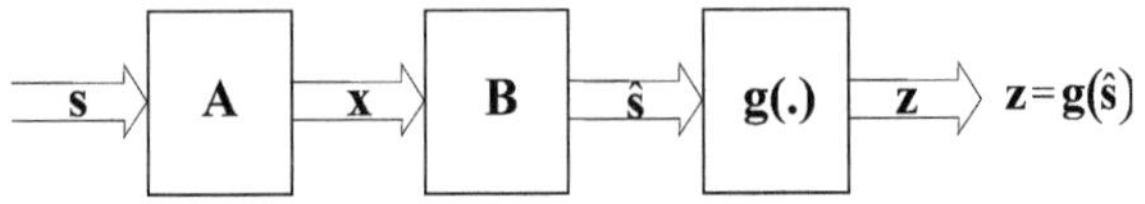

Fig. 2.2 InfoMax principle: mixing, unmixing, and nonlinear transformation

procedure of deterministic algorithms can exploit the algebraic structure of the matrices involved. The components are extracted using two methods. The first method consists of extracting sources source by source (deflation method), i.e., $\mathbf{s}_i = \mathbf{b}_i\mathbf{x}$; the second one consists of extracting all the sources simultaneously (symmetric method). The contrasts corresponding to these methods are called one-unit (one component) and multi-unit (several or all components) contrast functions.

We selected some of the most representative ICA algorithms (InfoMax [32, 23], JADE [33], FastIca [20, 22], and TDSEP [34]) derived from different perspectives of contrast design (entropy-, moment/cumulant-, and correlation-based methods). These algorithms will be used in comparisons with the techniques proposed in this work. A brief review of the selected ICA algorithms is included below.

2.2.1 InfoMax

The InfoMax algorithm was proposed in [32]. The InfoMax principle consists of maximizing the output entropy of a system $\mathbf{z} = \mathbf{g}(\hat{\mathbf{s}}) = \mathbf{g}(\mathbf{Bx})$ with respect to the demixing matrix $\mathbf{B}$, where $\mathbf{g}$ is a nonlinear transformation (see Fig. 2.2).

The system shown in Fig. 2.2 can be considered as a neural network. The goal is to obtain the ICA parameters for an efficient flow of the information in the neural network. This requires maximizing the mutual information between the $\mathbf{x}$ inputs and the $\hat{\mathbf{s}}$ outputs. It can be demonstrated that under no noise assumption, the maximization of this mutual information is equivalent to the maximization of the joint (output) entropy [35].

The transformation $\mathbf{g}(\cdot)$ is a $R^n \rightarrow R^n$ component-wise, non-linear function that operates on the sources estimated by the system linear part, i.e., $[\mathbf{g}(\hat{\mathbf{s}})]_i = g_i(\hat{\mathbf{s}})\ \ 1 \leq i \leq n$. Thus, the InfoMax contrast function is defined as

$$\phi_I(\mathbf{B}) = H(\mathbf{g}(\mathbf{Bx})) \tag{2.10}$$

where $H(\cdot)$ is the differential entropy. Scalar functions $g_i, \ldots, g_n$ are taken to be "squashing functions" that are capable of mapping a wide input domain to a narrow output domain (0, 1), and to be monotonously increasing. The entropy output entropy is estimated as [5]

$$H(\mathbf{g}(\mathbf{Bx})) = \sum_i E\left[\log g_i'\left(b_i^T\mathbf{x}\right)\right] + \log|\det\mathbf{B}| \tag{2.11}$$

This expression can be matched with the expression of the likelihood in Eq. (2.9). If the nonlinearities g_i are chosen as the cumulative distribution functions corresponding to the densities p_i, i.e., $g_i'(\cdot) = p_i(\cdot)$, the output entropy is equal to the likelihood. Thus, InfoMax is equivalent to maximum likelihood estimation (see for instance [5, 36]).

The first implementation of InfoMax [32] employed a stochastic gradient algorithm. Afterwards, the algorithm convergence was accelerated using natural gradient [37]. InfoMax was extended in [23] (Extended InfoMax) for blind separation of mixed signals with sub- and super-gaussian source distributions. The optimization procedure uses stability analysis [38] to switch between sub- and super-gaussian regimes. The following is the algorithm learning rule

$$\Delta\mathbf{B} \propto \left(\mathbf{I} - E\left[\mathbf{g}(\hat{\mathbf{s}})\hat{\mathbf{s}}^T\right]\right)\mathbf{B} \tag{2.12}$$

$g_i^+(\hat{\mathbf{s}}_i) = -2\tanh(\hat{\mathbf{s}}_i)$ is usually used as component-wise nonlinearity for super-gaussian components and $g_i^-(\hat{\mathbf{s}}_i) = \tanh(\hat{\mathbf{s}}_i) - \hat{\mathbf{s}}_i$ for sub-gaussian components.

2.2.2 JADE

Joint Approximate Diagonalization of Eigen-matrices (JADE) is an algorithm that belongs to an approach derived from the theory of higher order cumulants [39]. This approach has been called higher-order cumulant tensor because its implementation is based on tensor algebra. The idea is to represent the fourth-order cumulant statistics of the data by a "quadricovariance tensor" and to compute its "eigenmatrices" to yield the desired components [40]. The tensor algebra enables the manipulation of the multidimensional higher-order cumulant matrices.

It can be shown that the second and third cumulants $cum(x_i, x_j)$ and $cum(x_i, x_j, x_k)$ are equal to the second and third moments $E[x_i, x_j]$ and $E[x_i, x_j, x_k]$. However, the fourth cumulant differs from the fourth moment of the random variables x_i, x_j, x_k, and x_l; this is defined as

$$\begin{aligned} cum(x_i x_j x_k x_l) &= C_{ijkl}(x_i x_j x_k x_l) \\ &= E[x_i x_j x_k x_l] - E[x_i x_j]E[x_k x_l] - E[x_i x_k]E[x_j x_l] - E[x_i x_l]E[x_j x_k] \end{aligned} \tag{2.13}$$

For independent variables $cum(x_i, x_j, x_k, x_l) = 0$. It means that $C_{ij}(\mathbf{s}) = \sigma_i^2\delta_{ij}$, $C_{ijkl}(\mathbf{s}) = k_i\delta_{ijkl}$ with $\delta_{ij} = 1$ for $i = j$ and $\delta_{ij} = 0$ for $i \neq j$, $\delta_{ijkl} = 1$ for $i = j = k = l$ and $\delta_{ijkl} = 0$ for $i \neq j \neq k \neq l$; where σ_i^2 is the variance, and k_i is the kurtosis of the source component $s_i(\sigma_i^2 = E[s_i^2]$, $K_i = E[s_i^4] - 3E^2[s_i^2])$ [5].

Thus, a measure of distance between the estimated and the source components can be stated as a distance between cumulants, obtaining the contrast under the whitening constraint

$$-\sum_i k_i C_{iiii}(\hat{\mathbf{s}}) = -E\left[\sum_i k_i\left(\hat{s}_i^4 - 3\right)\right] \tag{2.14}$$

If there is no prior knowledge about the sources in this case about the kurtosis, the contrast function is $-\sum_i k_i C_{iiii}^2(\hat{\mathbf{s}})$. This is equivalent to $\sum_{ijkl \neq iiii} C_{ijkl}^2(\hat{\mathbf{s}})$ since $E[\hat{\mathbf{s}}\hat{\mathbf{s}}] = \mathbf{I}$ [17] (up to a constant).

The JADE algorithm [33] approximates the independence by minimizing a smaller number of cross cumulants

$$\phi_{JADE} = \sum_{ijkl \neq ijkk} C_{ijkl}^2(\hat{\mathbf{s}}) \tag{2.15}$$

The optimization procedure of JADE tries to find the rotation matrix $\mathbf{W}$ such that the cumulant matrices $\{\mathbf{Q}_i^z\}$ of the whitened data $\mathbf{z} = \mathbf{Vx}$ are as diagonal as possible. This solves

$$\arg\min \sum_i \mathit{off}\left(\mathbf{W}\mathbf{Q}_i^z\mathbf{W}^T\right) \tag{2.16}$$

where the operator $\mathit{off}(\mathbf{M}) = \sum_{i \neq j} \mathbf{M}_{i,j}^2$ is the sum of the square of the off-diagonal elements $\mathbf{M}$. This algorithm is based on the Jacobi method whose principle is that the rotation matrix $\mathbf{Q}$ can be approximated by a sequence of elementary rotations $T_k(\phi_k)$ each of which try to minimize the off diagonal elements of the respective cumulant matrices. The rotation angle ϕ_k (Givens angles) can be calculated in closed form because fourth-order contrasts are polynomial in the parameters [41]. The rotation uses a small angle $\theta_{\min}$, which controls the accuracy of the optimization. Thus, cumulant-based algebraic techniques avoid having to use gradient techniques for optimization. A comprehensive review about higher-order contrast used in ICA and comparison with gradient-based techniques is in [42].

2.2.3 *FastIca*

ICA methods have also been approached from the nongaussianity perspective. As stated above, without nongaussianity the estimation of the independent components is not possible. It is well-known from the central limit theorem that the distribution of a sum of independent random variables tends toward a Gaussian distribution, under certain conditions. The ICA estimation can be formulated as the search for directions that are maximally non-gaussian. Each local maximum gives one independent component [5]. In addition, the Gaussian variable has the maximum differential entropy (for unbounded variables with a common given variance). Thus, in order to find one independent component, we have to minimize entropy, i.e., we have to maximize the nongaussianity.

Two classical methods employed for measuring nongaussianity in ICA are kurtosis and neguentropy. The kurtosis (fourth-order cumulant) of a random variable u is defined by $k(u) = E[u^4] - 3E^2[u^2]$. It is zero for Gaussian random variables and non zero for non Gaussian distributions. Random variables with negative kurtosis are called sub-gaussian or platykurtic, (e.g., the uniform random variable); and those with positive kurtosis are called super-gaussian or leptokurtic (e.g., the Laplacian random variable). Thus, functions such as $-\sum_i |k(\hat{s}_i)|$ and $-\sum_i |k^2(\hat{s}_i)|$are appropriate contrasts. The gradient algorithm associated with the absolute value of the kurtosis is:

$$\Delta \mathbf{W} \alpha \, \text{sign}\left(\text{k}\left(\mathbf{w}^T \mathbf{z}\right)\right) E\left[\mathbf{z}\left(\mathbf{w}^T \mathbf{z}\right)^3\right] \tag{2.17}$$

with the projection of $\mathbf{W}$ on the unit sphere every step, i.e., it is normalized: $\frac{\mathbf{w}}{\| \mathbf{w} \|}$. This algorithm finds one component at a time, working with a whitened version of the mixed sources, $\mathbf{z} = \mathbf{Vx}$ by finding a column vector $\mathbf{W}$ that maximizes the module of the kurtosis of $\hat{\mathbf{s}} = \mathbf{w}^T \mathbf{z}$.

The FastIca algorithm uses estimates of neguentropy based on the maximum entropy principle, which requires the use of appropriate nonlinearities for the learning rule of the neural network [20, 22]. Separation is performed by the minimization of the neguentropy of the mixture in order to obtain uncorrelated and independent sources whose amplitude distributions are as non Gaussian as possible. The non Gaussianity is measured with the differential entropy j, called neguentropy [17], which is defined as the difference between the entropy of a Gaussian random variable u_{gauss} and the differential entropy of a random variable u, which are both variables of the same correlation (and covariance) matrix

$$J(u) = H\left(u_{gauss}\right) - H(u) \tag{2.18}$$

where the differential entropy H is defined by $H(u) = -\int f(u) \log f(u) du$. Since Gaussian random variables have the largest entropy H among all random variables having equal variance, maximizing $J(u)$ leads to the separation of independent source signals.

The use of neguentropy has an advantage that is well justified by statistical theory. However, entropy estimation is computationally very difficult. Thus, several methods of approximation have been proposed [5]. One successful approximation consists of using a nonquadratic function G, which becomes

$$J(u) \alpha [E\{G(u)\} - E\{G(v)\}]^2 \tag{2.19}$$

For optimization, the following algorithm can be obtained

$$\Delta \mathbf{w} \, \alpha \, \gamma E\left\{\mathbf{z} g\left(\mathbf{w}^T \mathbf{z}\right)\right\} \tag{2.20}$$

with the projection of $\mathbf{w}$ on the unit sphere every step and where $\gamma = E\{G(\mathbf{w}^T\mathbf{z})\} - E\{G(v)\}$ and v is a standardized Gaussian random variable. The normalization is necessary to project $\mathbf{w}$ to keep the variance of $\mathbf{w}^T\mathbf{z}$ constant. The non-linearity $g(\cdot)$ is the derivative of the function G used in the approximation. It can be chosen from $g_1(\hat{s}) = \tanh(a_1\hat{s})$ where $1 \le a_1 \le 2$, $g_2(\hat{s}) = \hat{s}\exp(-\hat{s}^2/2)$, or $g_3(\hat{s}) = \hat{s}^3$ [20, 22].

2.2.4 TDSEP

Temporal decorrelation source separation (TDSEP) is one of the ICA algorithms that exploit the time structure of the signals. It is based on the simultaneous diagonalization of several time-delayed correlation matrices. The approach relies on second-order statistics by assuming distinctive spectral/temporal characteristics of the sources [34, 43, 44]. These algorithms have been successfully applied in biosignal processing given the inherent time structure of the signals and their capability to separate signals whose amplitude distribution is near Gaussian.

The TDSEP algorithm uses the property that the cross-correlation functions vanish for mutually independent signals. It assumes that the signals $\mathbf{s}(t)$ have temporal structure ("non delta" autocorrelation function). All time delayed correlation matrices $\mathbf{R}_{\tau(\mathrm{s})}$ should be diagonal. This knowledge is used to calculate the unknown mixing matrix $\mathbf{A}$ by a simultaneous diagonalization of a set of correlated matrices $\mathbf{R}_{\tau(\mathrm{x})} = \langle \mathbf{x}(t)\mathbf{x}(t-\tau)^T \rangle$ for different choices of τ, where τ, is a lag constant, $\tau = 1, 2, 3, \ldots$. The diagonal elements of these matrices are formed by the values of the autocorrelation functions and the off-diagonal elements are the respective cross correlations,

$$R_{\tau(\mathbf{x})} = \begin{bmatrix} \varphi_{x_1,x_1} & (\tau)\ldots\varphi_{x_1,x_n} & (\tau) \\ \varphi_{x_1,x_2} & (\tau)\ldots\varphi_{x_2,x_n} & (\tau) \\ \vdots & \ddots & \vdots \\ \varphi_{x_n,x_1} & (\tau)\ldots\varphi_{x_n,x_n} & (\tau) \end{bmatrix} \tag{2.21}$$

where φ denotes the correlation function. If the signals were independent over time, all time-delayed correlation matrices should be diagonal because the cross-correlations of independent signals vanish.

The contrast consists of finding a matrix $\mathbf{B}$ (considering whitening) so that in addition to making the instantaneous covariances of $\hat{\mathbf{s}}(t) = \mathbf{B}\mathbf{x}(t)$ go to zero, the lagged covariances are made zero as well:

$$E\left[\hat{s}_i(t)\hat{s}_j(t-\tau)\right] = 0, \quad \text{for all } i, j, \tau \quad \text{with } i \neq j \tag{2.22}$$

For the independent components $S_i(t)$, the lagged covariances are all zero due to independence, without the need for higher-order information to estimate the model.

The optimization procedure has to minimize the sum of the off-diagonal elements (diagonalize) of several lagged covariances of $\hat{\mathbf{s}} = \mathbf{wz}$. Considering the symmetric version $\bar{\mathbf{C}}_{\tau(\mathbf{z})} = \frac{1}{2}\left[\mathbf{C}_{\tau(\mathbf{z})} + \left(\mathbf{C}_{\tau(\mathbf{z})}\right)\right]^T$ of the covariance matrix and a set of chosen lags τ denoted by **s**, the objective function can be written as

$$\sum_{\tau \in S} off\left(\mathbf{W}\bar{\mathbf{C}}_{\tau(\mathbf{z})}\mathbf{W}^T\right) \tag{2.23}$$

The minimization of Eq. (2.23) can be accomplished by a gradient descent algorithm. Another alternative is to adapt the existing methods for eigenvalue decomposition for this simultaneous approximate diagonalization of several matrices. The SOBI algorithm (Second-Order Blind Identification) and TDSEP use Jacobi-like algorithms for optimization [43, 44].

The set of time delays τ can be arbitrarily selected or manually given with prior knowledge. The advantage of second-order methods is their computational simplicity and efficiency. Furthermore, for a reliable estimate of covariances only comparatively few samples are needed.

2.3 Non-Parametric ICA

The estimation of the densities is, in general, a non-parametric problem. This means that the number of parameters is infinite, or, in practice, very large. The non-parametric problems are the most difficult to estimate. As was reviewed in Sect. 2.1, most known methods for solving the ICA problem involve specification of the parametric form of the latent components densities p_i and estimation of **B** together with parameters of p_i using maximum likelihood or minimization of the empirical versions of various divergence criteria between densities. In practical applications, the distributions p_i of the independent components are generally unknown, and thus ICA can be considered as a semi-parametric method in which these distributions are left unspecified.

Conventional ICA techniques have used two methods to avoid non-parametric estimation. The first method consists of using prior available knowledge about the densities. The results of the estimator would depend on the specification of the priors. By including these priors in the likelihood, the likelihood would really be a function of **B** only. A second method is to approximate the densities of the independent components by a family of densities that are specified by a limited number of parameters. For instance, a simple parameterization of the p_i is a single binary parameters, i.e., the choice between two densities [5].

Nowadays, there seem to be two research directions in ICA modelling: the first is motivated to design a signal separation algorithm that is "truly blind" to the particular underlying distributions of the mixed signals (any information about the sources is completely unknown), (see for instance [45]); the second consists of including the maximum number of priors available in the cost function in order to guide the algorithm to find particular sources (blind source extraction, semi-blind source separation, etc.), (see for instance [46, 47]). Some new methods that use non-parametric (NP) density estimation have been recently developed from the first direction in ICA research.

The new non-parametric ICA methods use techniques such as: minimization of a kernel canonical correlation or a kernel generalized variance among recovered sources (the so-called Kernel-ICA) [48]; maximum likelihood estimation (MLE) by using spline-based density approximations [49]; MLE by using Gaussian kernel density estimates (the so-called Npica) [45]; and minimization of the entropy of the marginals by estimating their order statistics (the so-called Radical) [50]. These methods have shown good performance in simulations, but there are no references about their performance in real applications. Theoretical analyses (convergence, consistency, and other issues) of non-parametric density estimation in the framework of ICA are found in [29, 26, 51]. We include a review of the Npica, Radical, and Kernel-ICA algorithms in the following section.

2.3.1 Npica

The Npica algorithm [45] is a maximum loglikelihood ICA method that solves the Eq. (2.9). It uses a non-parametric estimation for the probability density function p_i, which is directly estimated from the data using a kernel density estimation technique [52].

Given a batch of sample data of size N, the marginal distribution of an arbitrary reconstructed signal is approximated as follows:

$$p_i(\hat{s}_i) = \frac{1}{Nh}\sum_{l=1}^{N} \kappa\left(\frac{\hat{s}_i - \hat{s}_{il}}{h}\right), \quad i = 1,\ldots,M \tag{2.24}$$

where h is the kernel bandwidth and κ is the Gaussian kernel $\kappa(u) \triangleq \frac{1}{\sqrt{2\pi}} e^{-u^2/2}$.

The kernel centroids $\hat{\mathbf{s}}_{il}$ are equal to $\hat{\mathbf{s}}_{il} = \mathbf{w}_i\mathbf{x}^{(l)} = \sum_{l=1}^{N} w_{il}\mathrm{X}_{li}$, where $\mathbf{x}^{(l)}$ is the lth column of the mixture matrix $\mathbf{X}$.

The expectation of the maximum loglikelihood solution is approximated by the following cost function

$$L(\mathbf{W}) = -L_0(\mathbf{W}) - \log(\det \mathbf{W}) \tag{2.25}$$

where $L_0(\mathbf{W})$ is obtained by replacing the marginal pdf's p_i with their kernel density estimates

$$\begin{aligned} L_0(\mathbf{W}) &= \sum_{i=1}^{M} E \log\left[\frac{1}{Nh}\sum_{l=1}^{M}\kappa\left(\frac{\hat{s}_i - \hat{s}_{il}}{h}\right)\right] \\ &\approx \frac{1}{N}\sum_{i=1}^{M}\sum_{k=1}^{M}\log\left[\frac{1}{Nh}\sum_{l=1}^{M}\kappa\left(\frac{\mathbf{w}_i\left(\mathbf{x}^{(k)} - \mathbf{x}^{(l)}\right)}{h}\right)\right] \end{aligned} \tag{2.26}$$

The overall optimization problem can thus be posed as

$$\min_{\mathrm{W}} -\frac{1}{N}\sum_{i=1}^{M}\sum_{k=1}^{M}\log\left[\frac{1}{Nh}\sum_{l=1}^{M}\kappa\left(\frac{\mathbf{w}_i\left(\mathbf{x}^{(k)} - \mathbf{x}^{(l)}\right)}{h}\right)\right] - \log|\det \mathbf{W}| \tag{2.27}$$

$$\mathrm{s.t}\|\mathrm{w}_i\| = 1,\ i = 1,\ldots,M \tag{2.28}$$

Given the sample data $\mathbf{x}^{(k)},\ k = 1,\ldots,N$, the objective of Eq. (2.27) is a smooth nonlinear function of the elements of the matrix $\mathbf{W}$. The additional constraints of Eq. (2.28) restrict the space of possible solutions of the problem to a finite set. The optimization technique applied is the quasi-Newton method.

2.3.2 *Radical*

The Radical algorithm [50] uses entropy minimization, i.e., it must estimate the entropy of each marginal for each possible $\mathbf{W}$ matrix. The Radical marginal entropy estimates are functions of the order statistics of those marginals.

The order statistics are estimated using *spacings* estimates of entropy. Consider a one-dimensional random variable Z, and a random sample of Z denoted by $Z^1, Z^2,\ldots,Z^N$. The order statistics of a random sample of Z are simply the elements of the sample rearranged in non-decreasing order: $Z^{(1)} \leq Z^{(2)} \leq \cdots \leq Z^N$. A spacing of order m, or $m - spacings$ is then defined to be $Z^{(i+m)} - Z^N$, for $1 \leq i \leq i + m \leq N$. Finally, if m is a function of N, a $m_N - spacings$ such as $Z^{(i+m)} - Z^{(i)}$, can be defined.

For any random variable Z with an impulse-free density $p(\cdot)$ and continuous distribution function $p(\mathbf{x}/C_k) = \left|\det \mathbf{A}_k^{-1}\right| p(\mathbf{s}_k)$, the following holds. Let $p*$ be the Z-way product density $p*(Z^1, Z^2,\ldots,Z^N) = p(Z^1)p(Z^2)\ldots p(Z^N)$. Then

$$E_{p*}\left[P\left(Z^{(i+1)}\right) - P\left(Z^{(i)}\right)\right] = \frac{1}{N+1},\quad \forall i, 1 \leq i \leq N-1 \tag{2.29}$$

Using these ideas, the following simple entropy estimator can be derived.

$$\hat{H}_{m-spacings}\left(Z^1,\ldots,Z^N\right) \equiv \frac{m}{N-1}\sum_{i=0}^{\frac{N-1}{m}-1}\log\left(\frac{N+1}{m}\left(Z^{(m(i+1)+1)}-Z^{(mi+1)}\right)\right) \tag{2.30}$$

Under the condition that $m, N \to \infty$, $\frac{m}{N} \to 0$, this estimator is consistent; typically $m = \sqrt{N}$. The intuition behind this estimator is that by considering $m-spacings$ with larger and larger values of m, the variance of the probability mass of these spacings relative to their expected values gets smaller and smaller. In fact, the probability mass of $m-spacings$ is distributed according to a beta distribution with parameters m and $N+1$ [50]. Thus, a modification of Eq. (2.30) in which the $m-spacings$ overlap is used in Radical. The final contrast consists of an entropy estimator that is used to minimize Eq. (2.8),

$$\hat{H}_{Radical}\left(Z^1,\ldots,Z^N\right) \equiv \frac{1}{N-m}\sum_{i=1}^{N-m}\log\left(\frac{N+1}{m}\left(Z^{(i+m)}-Z^{(i)}\right)\right) \tag{2.31}$$

The optimization method of the algorithm for cost function minimization is exhaustive search. It is assumed that the data are first pre-whitened and augmented with a number of synthetic replicates of each of the original N sample points with additive spherical Gaussian noise to make a surrogate data set. This is done in order to obtain a smoother version of the estimator in an attempt to remove false minima. Afterwards, for each angle θ, the data are rotated ($\hat{\mathbf{s}} = \mathbf{W}(\theta)\cdot\mathbf{x}$) using a pair-wise Jacobi rotation and the cost function evaluated. The output is the $\mathbf{W}$ corresponding to the optimal θ. There are $M(M-1)/2$ distinct Jacobi rotations parameterized by θ (for a M-dimensional ICA). Optimizing over a set of these rotations is known as a sweep. Empirically, performing multiple sweeps improves the estimate of $\mathbf{W}$ for some number of iterations. In [50], good results were reported in simulations for S $\approx$ M (S is the number of sweeps).

2.3.3 Kernel-ICA

The Kernel-ICA algorithm [48] uses contrast functions based on canonical correlations in a reproducing kernel Hilbert space. This approach is not based on a single nonlinear function, but rather on an entire function space of candidate nonlinearities. The contrast function is a rather direct measure of the dependence of a set of random variables. Considering the case of two univariate random variables x_1 and x_2, and letting F be a vector space of functions from $\mathbb{R}$ to $\mathbb{R}$, the F - *correlation* ρF is defined as the maximal correlation between the random variables $f_1(x_1)$ and $f_1(x_1)$, where f_1 and f_2 range over F:

$$\rho_{\mathrm{F}} = \max_{f_1,f_2\in\mathrm{F}} \operatorname{corr}(f_1(x_1), f_2(x_2)) = \max_{f_1,f_2\in\mathrm{F}} \frac{\operatorname{cov}(f_1(x_1), f_2(x_2))}{(\operatorname{var} f_1(x_1))^{1/2}(\operatorname{var} f_2(x_2))^{1/2}} \tag{2.32}$$

Clearly, if the variables x_1 and x_2 are independent, then the F - *correlation* is equal to zero. It can be shown that F - *correlation* is the maximal possible correlation between one-dimensional linear projections $\Phi(x_1)$ and $\Phi(x_2)$, with $\Phi(x) = K(\cdot, x)$ being the feature map, where the kernel $K(\cdot, x)$ is a function in F for each x. This is the definition of the first "canonical correlation" between $\Phi(x_1)$ and$\Phi(x_2)$.

Canonical correlation analysis (CCA) is a multivariate statistical technique similar to PCA. While PCA works with a single random vector and maximizes the variance of projections of the data, CCA works with a pair of random vectors (or in general with a set of m random vectors) and maximizes correlation between sets of projections. While PCA leads to an eigenvector problem, CCA leads to a generalized eigenvector problem.

Kernel-ICA employs a "kernelized" version of CCA to compute a flexible contrast function for ICA. The following definitions are considered. Let $\{x_1^1, \ldots, x_1^N\}$ and $\{x_2^1, \ldots, x_2^N\}$ denote sets of N observations of x_1 and x_2, respectively, and let $\{\Phi(x_1^1), \ldots, \Phi(x_1^N)\}$ and $\{\Phi(x_2^1), \ldots, \Phi(x_2^N)\}$ denote the corresponding images in feature space. Let S_1 and S_2 represent the linear spaces spanned by the α_i-images of the data points. Thus, $f_1 = \sum_{k=1}^{N} \alpha_1^k \Phi(x_1^k) + f_1^{\perp}$ and $f_2 = \sum_{k=1}^{N} \alpha_2^k \Phi(x_2^k) + f_2^{\perp}$, where $f_1^{\perp}$ and $f_2^{\perp}$ are orthogonal to S_1 and S_2, respectively.

Considering that K_1 and K_2 are the Gram matrices associated with the data sets $\{x_1^i\}$ and $\{x_2^i\}$, respectively, the following variance estimates are obtained: $\hat{\text{var}}(\langle \Phi(x_1), f_1 \rangle) = \frac{1}{N} \alpha_1^T K_1 K_1 \alpha_1$ and $\hat{\text{var}}(\langle \Phi(x_2), f_2 \rangle) = \frac{1}{N} \alpha_2^T K_2 K_2 \alpha_2$. Thus, the kernelized CCA problem for two variables becomes that of performing the following maximization:

$$\rho_F(K_1, K_2) = \max_{\alpha_1, \alpha_2 \in \mathbb{R}^N} \frac{\alpha_1^T K_1 K_2 \alpha_2}{\left(\alpha_1^T K_1^2 \alpha_1\right)^{1/2} \left(\alpha_2^T K_2^2 \alpha_2\right)^{1/2}} \tag{2.33}$$

The formulation as a generalized and regularized eigenvalue problem to m variables is the following:

$$\begin{pmatrix} \left(K_1 + \frac{N_\kappa}{2} I\right)^2 & K_1 K_2 & \ldots & K_1 K_m \\ K_2 K_1 & \left(K_2 + \frac{N_\kappa}{2} I\right)^2 & \ldots & K_2 K_m \\ \vdots & \vdots & & \vdots \\ K_m K_1 & K_2 K_m & \ldots & \left(K_m + \frac{N_\kappa}{2} I\right)^2 \end{pmatrix} \begin{pmatrix} \alpha_1 \\ \alpha_2 \\ \vdots \\ \alpha_m \end{pmatrix}$$
$$= \lambda \begin{pmatrix} \left(K_1 + \frac{N_\kappa}{2} I\right)^2 & 0 & \ldots & 0 \\ 0 & \left(K_1 + \frac{N_\kappa}{2} I\right)^2 & \ldots & 0 \\ \vdots & \vdots & & \vdots \\ 0 & 0 & \ldots & \left(K_1 + \frac{N_\kappa}{2} I\right)^2 \end{pmatrix} \begin{pmatrix} \alpha_1 \\ \alpha_2 \\ \vdots \\ \alpha_m \end{pmatrix} \tag{2.34}$$

where N_κ is a small positive constant used for regularization. The minimal value of this problem is called the first kernel canonical correlation.

The Kernel-ICA algorithm proceeds as follows. Given a set of data vectors $\mathbf{x}^1, \mathbf{x}^2, \ldots, \mathbf{x}^N$, and given a parameter matrix W, we set $\hat{\mathbf{s}}^i = \mathbf{W}\mathbf{x}^i$, for each i, and thereby form a set of estimated source vectors $\hat{\mathbf{s}}^1, \hat{\mathbf{s}}^2, \ldots, \hat{\mathbf{s}}^N$. The m components of these vectors yield a set of m Gram matrices, $K_1, K_2, \ldots, K_m$, and these Gram matrices (which depend on $\mathbf{W}$) define the contrast function $C(\mathbf{W}) = I\lambda F\ (K1, \ldots, Km)$. The ICA algorithm minimizes this function with respect to $\mathbf{W}$.

The optimization technique used for Kernel-ICA is gradient descent (with line search) on an almost-everywhere differentiable function $C(\mathbf{W})$. The algorithm converges to a local minimum of $C(\mathbf{W})$ for any starting point. However, the ICA contrast functions have multiple local minima, and restarts are generally necessary if we are to find the global optimum. Empirically, the number of restarts was found to be small when the number of samples was sufficiently large so as to make the problem well-defined [48].

2.4 ICA Mixture Modelling

ICAMM is proposed in the framework of pattern recognition, considering that the observed data come from a mixture model and they can be categorized into several mutually exclusive classes. ICAMM assumes the underlying process that generated observed data is composed by multiple ICA models (data of each class are modelled as an ICA, i.e., linear combinations of independent non-gaussian sources). This modelling has been proposed in order to deal with the problems of the widely used mixture of Gaussians (MoG)-based modelling [53]. The principal limitations of MoG are: (i) the size (M^2) of each covariance matrix becomes extremely large when the dimension (M) of the problem increases; and (ii) each component is a Gaussian, which is a condition that is rarely found in real data sets. The antecedents of ICAMM can be found in [54] where each Gaussian of the mixture was replaced with a probabilistic principal component analysis (PPCA), allowing the covariance matrix dimension to be reduced, preserving the representation of the data. This PCA-based method was modified in [55] using variational Bayesian inference to infer the optimum number of analysers, obtaining the so-called Mixture of Factor Analysers. Afterwards, a robust approach for PPCA that exploits the adaptive distribution tails of the Student-t was proposed [56, 57]. This last allows the performance of the method is not spoiled by non-gaussian noise (e.g., outliers). Thus, ICA mixture modelling has been the natural evolution from these antecedents.

ICAMM was introduced in [58] considering a source model switching between Laplacian and bimodal densities. Afterwards, the model was extended using generalized exponential sources [59], self-similar areas such as mixtures of Gaussians sub-features using variational Bayesian inference [53], and sources with non-gaussian structures recovered by a learning algorithm using Beta divergence [56]. In addition, the automatic estimation of the number of ICA mixtures has been approached by variational Bayesian learning [60, 61] and on-line adaptive estimation of the clusters comparing log-likelihood of the data [62]. An alternative to the simultaneous estimation of all the ICAMM parameters is the performing of segmented and repeated ICAs. This strategy has been recently applied for the extraction of neural activity from large-scale optical recordings [63]. Ultimately, computational optimization of gradient techniques used in ICAMM algorithm was proposed applying Newton's method in the [64, 60].

The general formulation of ICAMM is:

$$\mathbf{x}_t = \mathbf{A}_k \mathbf{s}_k + \mathbf{b}_k, \quad k = 1, \ldots, K \tag{2.35}$$

where C_k denotes the class k, and each class is described by an ICA model with a mixing matrix $\mathbf{A}_k$, and a bias vector $\mathbf{b}_k$. Essentially, $\mathbf{b}_k$ determines the location of the cluster and $\mathbf{A}_k \mathbf{s}_k$ its shape. The goal of an ICA mixture model algorithm is to determine the parameters for each class. Figure 2.3 shows the model of ICA mixtures.

There are a few methods proposed in the ICAMM framework. They can be grouped as follows: maximum-likelihood based, iterative-based on a distance measure, and variational Bayesian learning methods. We include a review of three representative ICAMM techniques: the first proposed method for unsupervised classification and automatic context switching [58], the Beta-divergence method [65], and a variational Bayesian method [53].

2.4.1 Unsupervised Classification Using ICAMM

In [58], an unsupervised classification maximum-likelihood-based algorithm for modelling classes with non-gaussian densities (ICA structures) is proposed. The likelihood of the data is given by the joint density $p(\mathrm{X}|\Theta) = \prod_{i=1}^{T} p(\mathbf{x}_t|\Theta)$, with t being the data index $t = 1, \ldots, T$. The mixture density is $p(\mathbf{x}_t|\Theta) = \prod_{k=1}^{T} p(\mathbf{x}_t|C_k\theta_k)p(C_k)$, where $\Theta = (\theta_1, \ldots, \theta_k)$ are the unknown parameters for each of the component densities $p(\mathbf{x}|C_k, \theta_k)$, and C_k denotes the class k, $k = 1, \ldots, K$. The data within each class k are described by Eq. (2.35).

The log-likelihood of the data for each class is defined as

$$\log p(\mathbf{x}_t|C_k, \theta_k) = \log p(\mathbf{s}_k) - \log(\det|\mathbf{A}_k|) \tag{2.36}$$

and the probability for each class given the data vector $\mathbf{x}_t$ is: $p(C_k|\mathbf{x}_t, \Theta) = \frac{p(\mathbf{x}_t|\theta_k, C_k)p(C_k)}{\sum_{k=1}^{K} p(\mathbf{x}_t|\theta_k, C_k)p(C_k)}$.

Fig. 2.3 Outline of the ICA mixture model

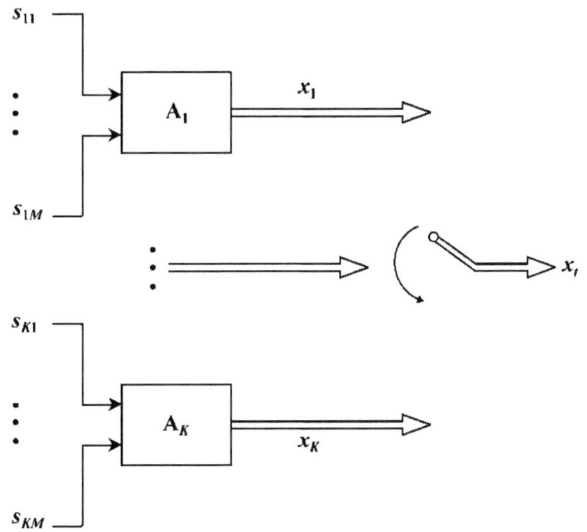

The Extended InfoMax algorithm [23] is used for adapting the basis functions (mixture matrix) in the ICA model. The gradient ascent technique is used to maximize the log-likelihood function. The rules to update the basis functions $\mathbf{A}_k$ and the bias vectors $\mathbf{b}_k$ for every class are the following

$$\Delta\mathbf{A}_k \propto -p(C_k|\mathbf{x}_t,\Theta)\mathbf{A}_k\left[\mathbf{I}-\mathbf{K}\ \tanh(\mathbf{s}_k)\mathbf{s}_k^T - \mathbf{s}_k\mathbf{s}_k^T\right] \tag{2.37}$$

$$\mathbf{b}_k = \frac{\sum_{t=1}^{T}\mathbf{x}_t p(C_k|\mathbf{x}_t,\Theta)}{\sum_{k=1}^{K} p(C_k|\mathbf{x}_t,\Theta)} \tag{2.38}$$

For the automatic switching between super-gaussian and sub-gaussian source distributions, a switching matrix $O_{k,l}$ is used. Super-Gaussian $\left(O_{k,l}=1\right):\log p(\mathbf{s}_k) \propto -\sum_{l=1}^{n}|s_{k,l}|$, and Sub-Gaussian $\left(O_{k,l}=-1\right):\log p(\mathbf{s}_k)\propto -\sum_{l=1}^{n}\left(\log\left(\cosh\left(s_{k,l}\right)\right) - \frac{s_{k,l}^2}{2}\right)$. where n is the dimensions of the source, $s_{k,l}$ is the lth dimension of the source in the kth class, and $O_{k,l}$ is an index which allows for automatic switching between super-gaussian and sub-gaussian models [23] $O_{k,l} = \mathrm{sign}\left[E\left\{\mathrm{sech}^2\left(s_{k,l}\right)\right\}E\left\{s_{k,l}^2\right\} - E\left\{\left(\tanh\left(s_{k,l}\right)\right)s_{k,l}\right\}\right]$.

The algorithm was tested to automatically identify different contexts in BSS (each context featured by the parameters of an ICA model), assuming the number of classes K to be known. An extension was made in [61] where the number of clusters and the intrinsic dimension of each cluster were determined by a variational Bayesian method similar to the method proposed in [59]. Recently, an on-line version for partitioning the input–output space for fuzzy neural networks was proposed in [62]. In this algorithm, one cluster is generated for the first data vector. For new data, a decision is made to generate or not generate new clusters depending on the degree to which the new incoming pattern $\mathbf{x}_t$belongs to the jth cluster, which is defined as $F^j(\mathbf{x}_t) = \log p(\mathbf{x}_t|C_j)$. The maximum log-likelihood value ($F^{J\max}(\mathbf{x}_t)$) among all log-likelihood values estimated for the existing J clusters at time t is selected. If $F^{J\max}(\mathbf{x}_t) \geq F$, the corresponding new incoming pattern is added to the existing cluster with index $J_{\max}$, and the parameters of this cluster are updated properly ($F$is a given negative threshold value obtained empirically). In this case, no new cluster is generated. If $F^{J\max}(\mathbf{x}_t) \geq F$, a new cluster is generated to accommodate this new pattern.

2.4.2 β-Divergence Method Applied to ICAMM

This algorithm is based on the minimum β-divergence distance [56, 65]. The β-divergence between two probability density functions $p(\mathbf{x})$ and $q(\mathbf{x})$ is defined as

$$D_\beta(p,q) = \int \left[\frac{1}{\beta}\{p^\beta(\mathbf{x}) - q^\beta(\mathbf{x})\}p(\mathbf{x}) - \frac{1}{\beta+1}\{p^{\beta+1}(\mathbf{x}) - q^{\beta+1}(\mathbf{x})\}\right] d\mathbf{x}, \quad \text{for } \beta > 0 \tag{2.39}$$

which is non-negative and equal to zero if and only if $p(\mathbf{x}) = q(\mathbf{x})$. The β-divergence reduces to Kullback–Leibler divergence when $\beta \to 0$.

There exists a matrix $\mathbf{W}$ and a shifting parameter vector $\boldsymbol{\mu}$ such that the components of $\hat{\mathbf{s}} = \mathbf{W}\mathbf{x} - \boldsymbol{\mu}$. Thus, the joint density of $\hat{\mathbf{s}}$ can be expressed as the product of marginal density functions $q_1, \ldots, q_m$ by $q(\hat{\mathbf{s}}) = \prod_{i=1}^{m} q_i(\hat{s}_i)$, and the joint density function of $\mathbf{x}$ can be expressed as $r(\mathbf{x}, \mathbf{W}, \boldsymbol{\mu}) = |\det(\mathrm{W})| \prod_{i=1}^{m} q_i(\mathbf{w}_i\mathbf{x} - \mu_i)$, where $\mathbf{W}_i$ is the ith row vector of $\mathbf{W}$, and μ_i is the ith component of $\boldsymbol{\mu}$.

The algorithm explores the recovering matrix of each class in the ICA mixture on the basis of the initial condition of a shifting parameter $\boldsymbol{\mu}$. If the initial value of the shifting parameter is close to the mean of the kth class, then the estimates for the recovering matrix $\mathbf{W}_k$ and the shifting parameter $\boldsymbol{\mu}_k$ can be obtained for this class by considering the data in other classes as outliers. Thus, $\{(\mathbf{W}_k, \boldsymbol{\mu}_k);\ k = 1, \ldots, c\}$ can be estimated by the repeated application of the β-divergence method to recover all hidden classes that are sequentially based on a rule for the step-by-step change of the shifting parameter $\boldsymbol{\mu}$. In order to create a rule for the sequential change of $\boldsymbol{\mu}$, the weight function ϕ is defined

$$\phi(\mathbf{x}, \mathbf{W}, \boldsymbol{\mu}) = \prod_{i=1}^{m} p_i^\beta(\mathbf{w}_i\mathrm{x} - \mu_i) \tag{2.40}$$

The minimum β-divergence method finds the minimizer of the empirical β-divergence $\widehat{D}_\beta(\tilde{r}, r_0(\cdot, \mathbf{W}, \boldsymbol{\mu}))$, where $\tilde{r}$ is the empirical distribution of $\mathbf{x}$, and r_0

corresponds to a nonlinearity with density p_i (e.g., $p_i(z) = c_2/\cosh(z)$ for super-gaussian signals) that allows switching between sub-gaussian and super-gaussian densities as in the Extended InfoMax algorithm [23]. This minimization is equivalent to maximizing the following quasi β-likelihood function:

$$L_\beta(\mathbf{w}, \boldsymbol{\mu}) = \frac{1}{n}\sum_{t-1}^{n} l_\beta(\mathbf{x}_t; \mathbf{W}, \boldsymbol{\mu}) \tag{2.41}$$

where $l_\beta(\mathbf{x}; \mathbf{w}, \boldsymbol{\mu}) = \begin{cases} \log(r_0(\mathbf{x}, \mathbf{w}, \boldsymbol{\mu})), & \text{for } \beta = 0 \\ \frac{1}{\beta} r_0^\beta(\mathbf{x}, \mathbf{w}, \boldsymbol{\mu}) - b_\beta(\mathbf{w}) - \frac{1-\beta}{\beta}, & \text{for } 0 < \beta < 1 \end{cases}$, and

$b_\beta(\mathbf{w}) = \frac{1}{\beta+1}\int r_0^{\beta+1}(\mathbf{x}, \mathbf{w}, \boldsymbol{\mu})d\mathbf{x} = \frac{|\det(\mathbf{w})|^\beta}{\beta+1}\int \prod_{i=1}^{m} p_i^{\beta+1}(z_i)d\mathbf{z}$

2.4.3 Variational Mixture of Bayesian ICAs

Bayesian inference and variational learning were introduced in the estimation of the ICAMM parameters in [53]. Mixture of Gaussians was used as source model. The generative model for a data vector **x** in this approach is shown in Fig. 2.4.

The probability of generating a data vector $\mathbf{x}^n$ from a C-component mixture model given assumptions $\mathcal{M}$ is:

$$p(\mathbf{x}^n|\mathcal{M}) = \sum_{c=1}^{C} p(c|\mathcal{M})_0 p(\mathbf{x}^n|\mathcal{M}_c, c) \tag{2.42}$$

A data vector is generated by choosing one of the C components stochastically under $p(c|\mathcal{M})_0$ and then drawing from $p(\mathbf{x}^n|\mathcal{M}_c, c)$; where $\mathcal{M} = \{\mathcal{M}_0, \mathcal{M}_1, \ldots, \mathcal{M}_c\}$ is the vector of component model assumptions, $\mathcal{M}_c$, and assumptions about the mixture process, $\mathcal{M}_0$. The assumptions represent everything that essentially defines the model (values of fixed parameters, model structure, details of the component switching method, any prior information, etc.).

The probability of observing data vector $\mathbf{x}^n$ under component cth ICA model ($\mathbf{x} = \mathbf{A}_c\mathbf{s}_c + \mathbf{y}_c + \mathbf{e}_c$, $\mathbf{s}_c$ are the sources of dimension L_c, $\mathbf{y}_c$ is an S-dimensional bias vector, and **x** is S-dimensional additive noise) is given by

$$p(\mathbf{x}^n|\theta_c, c) = \left(\frac{\lambda_c}{2\pi}\right)^{\frac{S}{2}} \exp[-E_c] \tag{2.43}$$

where $\theta_c = \{\mathbf{A}_c, \mathbf{s}_c^n \lambda_c\}$, $E_c = \frac{\lambda_c}{2}\left(\mathbf{x}_n - \mathbf{A}_c\mathbf{s}_c^n - \mathbf{y}_c\right)^T\left(\mathbf{x}_n - \mathbf{A}_c\mathbf{s}_c^n - \mathbf{y}_c\right)$, and λ_c is related with the variance of the noise considered zero-mean Gaussian and isotropic.

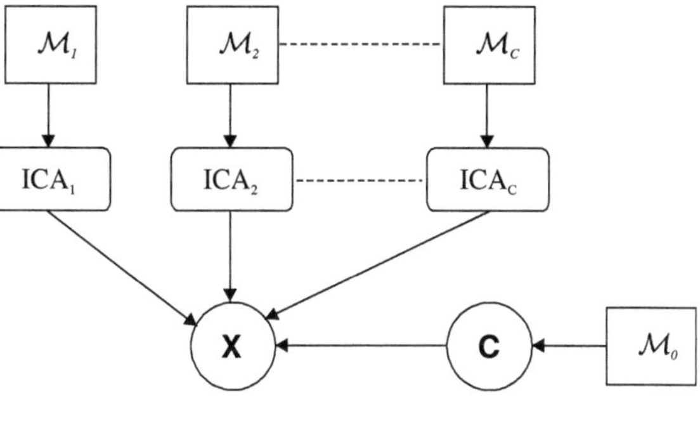

Fig. 2.4 ICA mixture for variational learning

The source model is MoG, which is a factorized mixture of 1-dimensional Gaussians with L_c factors (i.e., sources) and L_c components per source. This model is defined as (subscript c has been dropped for brevity),

$$
\begin{aligned}
p(\mathbf{s}_c^n|\varphi_c, c) &= \prod_{i=1}^{L_c}\sum_{q_i=1}^{m_i} p(q_i^n = q_i|\pi_i, c)p\left(\mathbf{s}_{c,i}^n|\varphi_{c,i}, c\right) \\
&= \prod_{i=1}^{L_c}\sum_{q_i=1}^{m_i} \pi_{i,q_i}\mathcal{N}\left(\mathbf{s}_{c,i}^n; \mu_{i,q_i}, \beta_{i,q_i}\right)
\end{aligned}
\tag{2.44}
$$

where μ_{i,q_i} is the position of feature q_i w.r.t. the cluster centre, β_{i,q_i} is its size, and π_{i,q_i} its "prominence" w.r.t. other features. The mixture proportions $\pi_{i,q_i} = p\left(q_1^n = q_i|\pi_i\right)$ are the prior probabilities of choosing component q_i of the ith source (of the cth ICA model etc.). q_i^n is a variable indicating which component of the ith source is chosen for generating $s_{c,i}^n$ and takes on values of $\{q_i = 1, \ldots, q_i = m_i\}$ (where m_i depends on ICA model c). The parameters of source i are $\varphi_{c,i} = \left\{\boldsymbol{\pi}_{c,i}, \boldsymbol{\mu}_{c,i}, \boldsymbol{\beta}_{c,i}\right\}$. The complete parameter set of the source model is $\varphi_c = \left\{\varphi_{c,1}, \varphi_{c,2}, \ldots, \varphi_{c,L_c}\right\}$. The complete collection of possible source states is denoted as $\mathbf{q}_c = \left\{\mathbf{q}_{c,1}, \mathbf{q}_{c,2}, \ldots \mathbf{q}_{c,\mathbf{m}}\right\}$ and runs over all $\mathbf{m} = \prod im_i$ possible combinations of source states.

It can be shown that the likelihood of the i.i.d. data $\mathbf{X} = \left\{\mathbf{x}^1, \mathbf{x}^2, \ldots \mathbf{x}^N\right\}$ given the model parameters $\Theta_c = \{\mathbf{A}_c, \mathbf{y}_c, \lambda_c, \ \varphi_c\}$ can be written as

$$
p(\mathbf{X}|\Theta_c, c) = \prod_{n=1}^{N}\sum_{\mathbf{q}=1}^{\mathbf{m}} \int p\left(\mathbf{x}^n, \mathbf{s}_c^n, \mathbf{q}_c^n|\Theta_c, c\right)d\mathbf{s}_c \tag{2.45}
$$

where $d\mathbf{s}_c = \prod ids_{c,i}$. Thus the probability of generating a data vector from a C-component mixture model can be written as

$$
p(\mathrm{X}|\mathcal{M}) = \sum_{c=1}^{C} p(c|\mathbf{k})p(\mathrm{x}|\Theta_c, c) \tag{2.46}
$$

where $p(c|\mathbf{k}) = \{p(c=1) = k_1, p(c=2) = k_2, \ldots, p(c=C) = k_c\}$. $p(\mathbf{x}|\mathcal{M})$ is known as the evidence for model $\mathcal{M}$ and quantifies the likelihood of the observed data under model $\mathcal{M}$. A Bayesian solution can be obtained by integrating out the parameters $\{\mathbf{k}, \Theta_c\}$ and hidden variables $\{\mathbf{s}_c, \mathbf{q}_c\}$. A set of prior distributions is assumed over all possible parameter values. For instance, the prior over the source model (MoG) parameters is defined as a product of priors over π_c, μ_c, β_c, thus $p(\varphi) = \prod_{c=1}^{C} p(\pi_c)p(\mu_c)p(\beta_c)$. In addition, the following priors are defined over: ICA mixture indicator variables $p(\mathbf{c}|\mathbf{k})$; ICA mixture coefficients $p(\mathbf{k})$; mixture proportions $p(\pi)$, mean and precision over each MoG $p(\boldsymbol{\mu})$ and $p(\beta)$; bias vector $p(\mathbf{y})$; sensor noise precision $p(\lambda)$; each element of the mixing matrix $p(\mathbf{A})$ with precision α_i for each column; and relevance of each source $p(\alpha)$.

The optimization follows from Bayes' rule $\log p(\mathbf{X}) = \log\frac{p(\mathbf{X},\psi)}{p(\psi|\mathbf{X})}$. The term ψ is the vector of all hidden variables and unknown parameters. This can be written as

$$\begin{aligned}\log p(\mathbf{X}) &= \int p'(\psi)\log\frac{p'(\psi)p(\mathbf{X},\psi)}{p'(\psi)p(\psi|\mathbf{X})}d\psi \\ &= \int p'(\psi)\log\frac{p'(\mathbf{X},\psi)}{p'(\psi)}d\psi + \int p'(\psi)\log\frac{p'(\psi)}{p'(\psi|\mathbf{X})}d\psi \\ &= \mathrm{F}[\psi] + KL[p'||p]\end{aligned} \tag{2.47}$$

where $p'(\psi)$ is some approximation to the posterior $p'(\psi|\mathbf{X})$; $F[\psi] = \langle\log\, p(\mathbf{X},\psi)\rangle_{p'(\psi)} + \mathcal{H}[p'(\psi)]$; and $KL[p'||p] = \int p'(\psi)\log\frac{p'(\psi)}{p(\psi|\mathbf{X})}d\psi$. $\mathcal{H}[p'(\psi)]$ is the entropy of $p'(\psi)$, and KL is the Kullback–Leibler divergence.

In the mixture model $\boldsymbol{\pi} = \{\mathbf{c}, \mathbf{s}, \mathbf{q}, \mathbf{k}, \boldsymbol{\Theta}\}$. By choosing $p'(\psi)$ such that it factorizes, terms in each hidden variable can be maximized individually. In [53], the following factorization was chosen,

$$p'(\psi) = p'(\mathbf{c})p'(\mathbf{s}_c|\mathbf{q}_c, c)p'(\mathbf{q}_c|c)p'(\mathbf{k})p'(\mathbf{y})p'(\lambda)p'(\mathbf{A})p'(\alpha)p'(\boldsymbol{\phi}) \tag{2.48}$$

where $p'(\boldsymbol{\phi}) = p'(\pi)p'(\boldsymbol{\mu})\mathbf{p}'(\boldsymbol{\beta})$ and $p'(a|b)$ is the approximating density of $p(a|b,\mathbf{X})$. Also the posteriors over the sources were factorized such that $p'(\mathbf{s}_c, \mathbf{q}_c|c) = \prod_{i=1}^{L_c} p'(q_c|c)p'(s_{c,i}|q_i, c)$.

2.5 Conclusions

In this chapter, an overview of the current techniques in ICA and ICA mixture modelling (ICAMM) has been carried out. These techniques establish a framework for non-linear processing of data with complex non-gaussian distributions.

Classical statistical signal processing relies on exploiting second-order information. Spectral analysis and linear adaptive filtering are probably the most representative examples. From the perspective of optimality (optimum detection and estimation), second-order statistics are sufficient statistics when Gaussianity holds, but lead to suboptimum solutions when dealing with general probability density models. A natural evolution of statistical signal processing, in connection with the progressive increase in computational power, has been exploiting higher-order information. Thus, high-order spectral analysis and nonlinear adaptive filtering have received the attention of many researchers in this field.

Clearly, within this framework of evolution from second-order to higher-order information, is the transition from PCA to ICA. Briefly, PCA is a technique for linearly transforming a vector of correlated components into a vector of variance-ordered uncorrelated components; meanwhile ICA linearly transforms a vector of statistically dependent components into unordered independent components. ICA can also be considered as a natural evolution of prewhitening linear transformation (like PCA but no variance ordering is being produced). When Gaussianity holds, both ICA and prewhitening get equivalent transformations, and infinite solutions may exist, as any rotation of the prewhitened vector keeps the uncorrelation among the vector components. However, when non-gaussianity appears, ICA produces a different transformation, which can be unique if appropriate constraints are introduced into the design. That is the reason why ICA has become so popular as a technique for blind source separation when at maximum one source is Gaussian.

Even more interesting is to recognize that ICA implicitly assumes a model for multivariate pdf's. The multivariate pdf of the transformed vector will be the product of the (one-dimensional) marginal pdf's of its components. Dealing with one-dimensional pdf's makes different complex problems involving multivariate pdf's tractable. This perspective suggests that ICA can be an interesting tool for use in areas of intensive data analysis. Actually, dealing with estimates of pdf's or defining optimality criteria involving pdf's (like entropy, mutual information, Kullback–Leibler distances, etc.) can be considered the last generation in statistical signal processing approaches: a natural evolution from second-order and higher-order statistics to data distribution information. In this chapter, we have reviewed some of the most representative ICA algorithms derived from entropy, cumulant, and time structure perspectives: InfoMax, JADE, FastIca, and TDSEP. In addition, we have reviewed the principal non-parametric ICA algorithms (Npica, Radical, Kernel-ICA) from a research direction that pursues generalization of the methods; and thus BSS is done with completely unknown information about the sources.

Some authors have termed the approaches above as non-linear information processing [66]. This is relevant since non-linear information processing establishes a bridge between statistical signal processing and computational and artificial intelligence sciences. That is why many people from signal processing are increasingly involved in areas like data mining, machine learning, or clustering, and many researchers from computational sciences are working on new data intensive signal and image processing applications.

Recently, ICAMM was introduced as an extension of ICA. ICAMM is a kind of nonlinear ICA technique that extends the linear ICA method by learning multiple ICA models and weighting them in a probabilistic manner. Thus, ICAMM has emerged as a flexible approach to model arbitrary data densities using mixtures of multiple ICA models with non-gaussian distributions for the independent components (i.e., relaxing the restriction of modelling every component by a multivariate Gaussian probability density function).

In this chapter, we reviewed three ICAMM methods. The first method is maximum likelihood-based, which uses the learning rule of extended InfoMax algorithm in the ICAMM parameter updating to distinguish between sub-gaussian and super-gaussian sources. The second method extracts the ICA classes sequentially from an initial estimate for each centroid and is based on a distance called Beta divergence, which is an extension of the Kullback–Leibler divergence. This method requires that various parameters be initialized, such as the beta value, initial centroids, a percentage of classification used as stopping criterion. These parameters are estimated rather arbitrarily. As in the first method, the extended InfoMax rule is used for unknown source distributions. Thus, the source model of these methods could only switch between Laplacian and bimodal densities, which is a limitation for source density estimation.

The third reviewed method is a variational Bayesian learning algorithm. This method uses a source model based on mixtures of Gaussians. In order to apply Bayesian inference, a set of prior distributions over all possible parameter values is assumed: source model (MoG), ICA mixture indicator variables; ICA mixture coefficients; mixture proportions, mean and precision over each MoG; bias vector; sensor noise precision; precision for each column of the mixing matrix; and relevance of each source. The algorithm uses variational optimization (to lighten the computationally expensive cost of Bayesian inference) to approximate integrating out the parameters. Taking into account that the source model for every ICA is MoG, the final ICAMM data model for this method can be also considered a kind of MoG.

In the case of the first two ICAMM methods reviewed, the number of clusters is known a priori. In the third method, the number of clusters can be estimated, although substantial a priori knowledge is required for the model parameters. All three methods consider only unsupervised learning; therefore, semi-supervised and supervised learning are left unspecified.

References

1. C. Jutten, J. Herault, Une solution neuromimétique au problème de séparation de sources. Traitement du Signal **5**(6), 389–404 (1989)
2. C. Jutten, J. Herault, Blind separation of sources, part I: an adaptive algorithm based on neuromimetic architecture. Signal Process. **24**, 1–10 (1991)
3. C. Jutten, J. Herault, Blind separation of sources, part II: problems statement. Signal Process. **24**, 11–20 (1991)

4. C. Jutten, J. Herault, Blind separation of sources, part III: stability analysis. Signal Process. **24**, 21–29 (1991)
5. A. Hyvärinen, J. Karhunen, E. Oja, *Independent Component Analysis* (Wiley, New York, 2001)
6. C.W. Hesse, C.J. James, On semi-blind source separation using spatial constraints with applications in EEG Analysis. IEEE Trans. Biomed. Eng. **53**(12-1), 2525–2534 (2006)
7. J. Even, K. Sugimoto, An ICA approach to semi-blind identification of strictly proper systems based on interactor polynomial matrix. Int. J. Robust Nonlinear Control **17**, 752–768 (2007)
8. Z. Ding, T. Ratnarajah, C.F.N. Cowan, HOS-based semi-blind spatial equalization for MIMO rayleigh fading channels. IEEE Trans. Signal Process. **56**(1), 248–255 (2008)
9. A. Cichocki, S. Amari, *Adaptive Blind Signal and Image Processing: Learning Algorithms and Applications* (Wiley, New York, 2001)
10. T.W. Lee, *Independent Component Analysis—Theory and Applications* (Kluwer Academic Publishers, Boston, 1998)
11. S. Roberts, R. Everson, *Independent Component Analysis—Principles and Practice* (Cambridge University Press, Cambridge, 2001)
12. A. Cichocki, R. Zdunek, A.H. Phan, S. Amari, *Nonnegative Matrix and Tensor Factorizations: Applications to Exploratory Multi-way Data Analysis and Blind Source Separation* (Wiley, Hoboken, 2009)
13. P. Comon, C. Jutten (eds.), *Handbook of Blind Source Separation Independent Component Analysis and Applications* (Academic Press, Oxford, 2010)
14. M.S. Pedersen, J. Larsen, U. Kjems, L.C. Parra, *A Survey of Convolutive Blind Source Separation Methods*, ed. by J. Benesty, A. Huang. Multichannel Speech Processing Handbook, Chapter 51 (Springer, Berlin, 2007), pp. 1065–1084
15. H. Buchner, R. Aichner, W. Kellerman, TRINICON: a versatile framework for multichannel blind signal processing. in *Proceedings of 29th IEEE International Conference on Acoustics, Speech, and Signal Processing, ICASSP*, pp. III-889–892, Montreal, Canada, 2004
16. W. Kellerman, H. Buchner, R. Aichner, Separating convolutive mixture with TRINICON. in *Proceedings of 31st IEEE International Conference on Acoustics, Speech, and Signal Processing, ICASSP*, pp. V-961–964, Toulouse, France, 2006
17. P. Comon, Independent component analysis—a new concept? Signal Process. **36**(3), 287–314 (1994)
18. S. Amari, A. Cichocki, H. Yang, *A new learning algorithm for blind signal separation, Advances in Neural Information Processing Systems*, vol 8 (MIT Press, Cambridge, 1996), pp. 752–763
19. S. Amari, J.F. Cardoso, Blind source separation-semiparametric statistical approach. IEEE Trans. Signal Process. **45**(11), 2692–2700 (1997)
20. A. Hyvärinen, E. Oja, A fast fixed-point algorithm for independent component analysis. Neural Comput. **9**(7), 1483–1492 (1998)
21. D.T. Pham, P. Garrat, Blind separation of mixture of independent sources through a quasi-maximum likelihood approach. IEEE Trans. Signal Process. **45**(7), 1712–1725 (1997)
22. A. Hyvarinen, Fast and robust fixed-point algorithms for independent component analysis. IEEE Trans. Neural Netw. **10**(3), 626–634 (1999)
23. T.W. Lee, M. Girolami, T.J. Sejnowski, Independent component analysis using an extended InfoMax algorithm for mixed sub-gaussian and super-gaussian sources. Neural Comput. **11**(2), 417–441 (1999)
24. S.I. Amari, T.P. Chen, A. Cichocki, Nonholonomic orthogonal learning algorithms for blind source separation. Neural Comput. **12**, 1463–1484 (2000)
25. J.F. Cardoso, Dependence, correlation and gaussianity in independent component analysis. J. Mach. Learn. Res. **4**, 1177–1203 (2003)
26. A. Chen, P.J. Bickel, Consistent independent component analysis and prewhitening. IEEE Trans. Signal Process. **53**(10), 3625–3632 (2005)
27. W. Liu, D.P. Mandic, A. Cichocki, Blind source extraction based on a linear predictor. IET Signal Process. **1**(1), 29–34 (2007)

28. J.F. Cardoso, Blind signal separation: statistical principles. *Proceedings of the IEEE. Special Issue on Blind Identification and Estimation*, vol 9, pp. 2009–2025, 1998
29. A. Chen, P.J. Bickel, Efficient independent component analysis. Annals Stat. **34**(6), 2825–2855 (2006)
30. F. Meinecke, A. Ziehe, M. Kawanabe, K.R. Müller, Resampling approach to estimate the stability of one-dimensional or multidimensional independent components. IEEE Trans. Biomed. Eng. **49**(12), 1514–1525 (2002)
31. J. Himberg, A. Hyvärinen, F. Esposito, Validating the independent components of neuroimaging time-series via clustering and visualization. Neuroimage **22**(3), 1214–1222 (2004)
32. A.J. Bell, T.J. Sejnowski, An information-maximization approach to blind separation and blind deconvolution. Neural Comput. **7**, 1129–1159 (1995)
33. J.F. Cardoso, A. Souloumiac, Blind beamforming for non gaussian signals. IEE Proc.-F **140**(6), 362–370 (1993)
34. A. Ziehe, K.R. Müller, TDSEP- an efficient algorithm for blind separation using time structure. *Proceedings of the 8th International Conference on Artificial Neural Networks, ICANN'98, Perspectives in Neural Computing*, pp. 675–680, 1998
35. J.P. Nadal, N. Parga, Non linear neurons in the noise limit: a factorial code maximizes information transfer. Netw. Comput. Neural Syst. **5**(3), 565–585 (1994)
36. J.F. Cardoso, InfoMax and maximum likelihood for blind source separation. IEEE Signal Process. Lett. **4**(4), 112–114 (1997)
37. S.I. Amari, Natural gradient works efficiently in learning. Neural Comput. **10**, 251–276 (1998)
38. J.F. Cardoso, B. Laheld, Equivariant adaptive source separation. IEEE Trans. Signal Process. **45**(2), 434–444 (1996)
39. C. Nikias, A. Petropulu, *Higher-order Spectral Analysis—A Nonlinear Signal Processing Framework* (Prentice Hall, Englewood Cliffs, 1993)
40. J.F. Cardoso, P. Comon, Tensor-based independent component analysis. *Proceedings of the Fifth European Signal Processing Conference, EUSIPCO 1990*, pp. 673–676, 1990
41. J.F. Cardoso, A. Souloumiac, Jacobi angles for simultaneous diagonalization. SIAM J. Matrix Anal. Appl. **17**(1), 161–164 (1996)
42. J.F. Cardoso, High-order contrasts for independent component analysis. Neural Comput. **11**(1), 157–192 (1999)
43. A. Ziehe, K.R. Muller, G. Nolte, B.M. Mackert, G. Curio, Artifact reduction in magnetoneurography based on time-delayed second order correlations. IEEE Trans. Biomed. Eng. **41**, 75–87 (2000)
44. A. Belouchrani, K. Abed-Meraim, J.F. Cardoso, E. Moulines, A blind source separation technique using second-order statistics. IEEE Trans. Signal Process. **45**, 434–444 (1997)
45. R. Boscolo, H. Pan, Independent component analysis based on nonparametric density estimation. IEEE Trans. Neural Netw. **15**(1), 55–65 (2004)
46. R. Boustany, J. Antoni, Blind extraction of a cyclostationary signal using reduced-rank cyclic regression—a unifying approach. Mech. Syst. Signal Process. **22**, 520–541 (2008)
47. J. Even, K. Sugimoto, An ICA approach to semi-blind identification of strictly proper systems based on interactor polynomial matrix. Int. J. Robust Nonlinear Control **17**, 752–768 (2007)
48. F.R. Bach, M.I. Jordan, Kernel independent component analysis. J. Mach. Learn. Res. **3**, 1–48 (2002)
49. T. Hastie, R. Tibshirani, Independent Component Analysis Through Product Density Estimation, Technical Report, Stanford University, 2002
50. E.G. Learned-Miller, J.W. Fisher, ICA using spacings estimates of entropy. J. Mach. Learn. Res. **4**, 1271–1295 (2003)
51. A. Samarov, A. Tsybakov, Nonparametric independent component analysis. Bernoulli **10**(4), 565–582 (2004)
52. B.W. Silverman, *Density Estimation for Statistics and Data Analysis* (Chapman and Hall, London, 1985)

53. R. Choudrey, S. Roberts, Variational mixture of bayesian independent component analysers. Neural Comput. **15**(1), 213–252 (2002)
54. M.E. Tipping, C.M. Bishop, Mixtures of probabilistic principal component analyzers. Neural Comput. **11**(2), 443–482 (1999)
55. Z. Ghahramani, M. Beal, Variational inference for Bayesian mixtures of factor analysers. Adv. Neural Inf. Process. Syst. **12**, 449–445 (2000)
56. C. Archambeau, N. Delannay, M. Verleysen, Mixtures of robust probabilistic principal component analyzers. Neurocomputing **71**(7–9), 1274–1282 (2008)
57. M. Svensén, C.M. Bishop, Robust Bayesian mixture modelling. Neurocomputing **64**, 235–252 (2005)
58. T.W. Lee, M.S. Lewicki, T.J. Sejnowski, ICA mixture models for unsupervised classification of non-gaussian classes and automatic context switching in blind signal separation. IEEE Trans. Pattern Anal. Mach. Intell. **22**(10), 1078–1089 (2000)
59. S. Roberts, W.D. Penny, Mixtures of independent component analyzers. in *Proceedings of ICANN2001*, Vienna, August 2001, pp. 527–534
60. J.A. Palmer, K. Kreutz-Delgado, S. Makeig, An Independent Component Analysis Mixture Model with Adaptive Source Densities, Technical Report, UCSD, 2006
61. K. Chan, T.W. Lee, T.J. Sejnowski, Variational learning of clusters of undercomplete nonsymmetric independent components. J. Mach. Learn. Res. **3**, 99–114 (2002)
62. C.T. Lin, W.C. Cheng, S.F. Liang, An on-line ICA-mixture-model-based self-constructing fuzzy neural network. IEEE Trans. Circuits Syst. **52**(1), 207–221 (2005)
63. T. Yoshida, M. Sakagami, K. Yamazaki, T. Katura, M. Iwamoto, N. Tanaka, Extraction of neural activity from in vivo optical recordings using multiple independent component analysis. IEEJ Trans. Electron. Inf. Syst. **127**(10), 1642–1650 (2007)
64. J.A. Palmer, S. Makeig, K. Kreutz-Delgado, B.D. Rao, Newton method for the ICA mixture model. *Proceedings of the 33rd IEEE International Conference on Acoustics, Speech, and Signal*, pp. 1805–1808, Las Vegas, USA, 2008
65. N.H. Mollah, M. Minami, S. Eguchi, Exploring latent structure of mixture ICA models by the minimum ß-Divergence method. Neural Comput. **18**, 166–190 (2005)
66. D. Erdogmus, J.C. Principe, From linear adaptive filtering to nonlinear information processing—the design and analysis of information processing systems. IEEE Signal Process. Mag. **23**(6), 14–33 (2006)

Chapter 3
Learning Mixtures of Independent Component Analysers

This chapter presents a new procedure for learning mixtures of independent component analyzers. The procedure includes non-parametric estimation of the source densities, supervised–unsupervised learning of the model parameters, incorporation of any independent component analysis (ICA) algorithm into the learning of the ICA mixtures, and estimation of residual dependencies after training for correction of the posterior probability of every class to the testing observation vector. We demonstrate the performance of the procedure in the classification of ICA mixtures of two, three, and four classes of synthetic data. The utilization of the proposed posterior probability correction demonstrates an improvement in the classification accuracy. Semi-supervised learning shows that unlabelled data can degrade the performance of the classifier when they do not fit the generative model. Comparative results of the proposed method and standard ICA algorithms for blind source separation in one and multiple ICA data mixtures show the suitability of the non-parametric ICA mixture-based method for data modelling.

Let us discuss a possible comparison of our method with other ICAMM methods such as those explained in Chap. 2. Note that the main objective of our approach is to pursue generalization in the ICAMM method. The proposed method is a maximum-likelihood approach; therefore, it can be related to the first method proposed for ICAMM by Lee et al. [1], which is also based on maximum-likelihood estimation. There are some significant differences in the proposed method that outperform Lee's method: (i) non-parametric source density estimation allows a wider range of densities to be modelled (e.g., complex multinomial densities), instead of simple switching between Laplacian and bimodal densities; (ii) different kinds of supervision in learning are allowed (unsupervised, semi-supervised, and supervised learning) compared with only the one kind of learning supported by Lee's method; (iii) the proposed method allows correction of residual dependencies after the learning stage; and (iv) incorporation of different methods for ICAMM parameter updating that are not supported by Lee's method.

A. Salazar, *On Statistical Pattern Recognition in Independent Component Analysis Mixture Modelling*, Springer Theses, DOI: 10.1007/978-3-642-30752-2_3,

The method based on Beta divergence proposed by Mollah et al. [2] is developed from a different approach than the one proposed. Mollah's method is a sequential method that extracts one ICA parameter set at a time while the proposed method is a block extraction technique that estimates the parameters for each ICA. The differences between the proposed method and Lee's method are also applicable for Mollah's method. In addition, Mollah's method requires application of various parameters (beta value, classification percentage for stopping criterion, etc.), which are estimated rather arbitrarily, making the method unstable. Therefore, it follows that the convergence to the optimum is not guaranteed.

The method proposed by Choudrey and Roberts [3] is based on variational Bayesian inference. This method defines priors over all possible parameters (ICA mixture indicator variables, ICA mixture coefficients, mixture proportions, mean and precision over each MoG, bias vector, etc.) of the ICA mixture. This kind of heavy parameterization requires great knowledge about the data. Broad priors have little or no effect on the results, but they can reduce their regularizing abilities leading to implausible magnitudes and possible over-fitting. Narrow priors encode strong assumptions about possible parameter and variable values; the model becomes inflexible and its ability to learn is compromised [3]. Thus, finding the right priors requires a great examination consisting in searching for an extensive number of parameters in a wide range of values. Another relevant feature of Choudrey and Roberts's method is that the source model for every ICA is MoG, and thus the final ICAMM data model is a kind of MoG. Choudrey and Roberts's method search for generalization, but tuning of the extensive number of prior parameters seems to be at best complicated. In contrast, the proposed method attempts to find a balance between parametric and non-parametric estimation. Thus, the ICA mixtures are modelled by a short set of parameters maintaining simplicity, but the source density estimation is flexible since it is non-parametric. In addition, the non-gaussianity of the data is preserved since any assumption about the source model is not imposed. This modelling is especially plausible in semi-supervised scenarios where fragmented knowledge is available as is common in real-world problems. With regard to the automatic estimation of the number of clusters, we can deal with this issue using the hierarchical algorithm proposed in Chap. 4. By defining a high number of clusters (ICA mixtures) at the bottom level of the hierarchy, an optimum number for clustering can be estimated at intermediate hierarchy levels using some stopping rules or cluster validation techniques.

From the above discussion, it is clear that are significant differences in the comparison of the previous methods with the proposed one. A competitive comparison of the proposed method with the previous methods by processing different cases is always desirable. However, since the implementation or adaptation of the previous methods would require great effort and the preparation of testing data sets was not straightforward considering the different perspectives from which the previous methods were made, we decided to study in greater depth our ICAMM method and apply it in different areas, instead of trying to compare it with other methods. This decision also provided the opportunity to apply our method to novel applications involving real-world problems.

3.1 The Model and the Definition of the Problem

In ICA mixture modelling, it is assumed that feature (observation) vectors $\mathbf{x}_k$ corresponding to a given class C_k $(k = 1\ldots K)$ are the result of applying a linear transformation $\mathbf{A}_k$ to a (source) vector $\mathbf{s}_k$, whose elements are independent random variables, plus a bias vector $\mathbf{b}_k$, i.e.,

$$\mathbf{x}_k = \mathbf{A}_k\mathbf{s}_k + \mathbf{b}_k \quad k = 1\ldots K \tag{3.1}$$

We assume that $\mathbf{A}_k$ is a square matrix: feature and source vectors have the same dimension. This is a practical assumption since original feature vectors are normally subjected to PCA and only the main (uncorrelated) components are retained for ICA, with no further reduction in the dimension of the new feature vector that is obtained in this way. An optimum classification of a given feature vector $\mathbf{X}$ of unknown class is made by selecting the class C_k that has the maximum conditional probability $p(C_k/\mathbf{x})$. Considering Bayes theorem, we can write:

$$p(C_k/\mathbf{x}) = \frac{p(\mathbf{x}/C_k)\cdot P(C_k)}{p(\mathbf{x})} = \frac{p(\mathbf{x}/C_k)\cdot p(C_k)}{\sum_{k'=1}^{K} p(\mathbf{x}/C_{k'})p(C_{k'})} \tag{3.2}$$

On the other hand, noting Eq. (3.1), if $\mathbf{x}$ is a feature vector corresponding to class C_k, then [4]

$$p(\mathbf{x}/C_k) = \left|\det \mathbf{A}_k^{-1}\right| p(\mathbf{s}_k) \tag{3.3}$$

where $\mathbf{s}_k = \mathbf{A}_k^{-1}(\mathbf{x} - \mathbf{b}_k)$. Considering Eqs. (3.1) and (3.2), we can write

$$p(C_k/\mathbf{x}) = \frac{\left|\det \mathbf{A}_k^{-1}\right| p(\mathbf{s}_k)p(C_k)}{\sum_{k'=1}^{K} \left|\det \mathbf{A}_{k'}^{-1}\right| p(\mathbf{s}_{k'})p(C_k)} \tag{3.4}$$

In conclusion, given a feature vector $\mathbf{x}$, we should compute the corresponding source vectors $\mathbf{s}_k\, k = 1\ldots K$, from $\mathbf{s}_k = \mathbf{A}_k^{-1}(\mathbf{x} - \mathbf{b}_k)$ to finally select the class having the maximum computed value $\left|\det \mathbf{A}_k^{-1}\right| p(\mathbf{s}_k)p(C_k)$ (Note that the denominator in Eq. (3.4) does not depend on k, so it does not influence the maximization of $p(C_k/\mathbf{x})$). To make the above computation, we need to estimate $\mathbf{A}_k^{-1}$, $\mathbf{b}_k$ (to compute $\mathbf{s}_k$ from $\mathbf{x}$) and the multidimensional pdf of the source vectors for every class (to compute $p(\mathbf{s}_k)$). Two assumptions are considered to solve this problem. First, that the elements of $\mathbf{s}_k$ are independent random variables (ICA assumption), so that the multidimensional pdf can be factored into the corresponding marginal pdf's of every vector element. The second assumption is that there is a set of independent feature vectors (learning vectors) available, which are represented by matrix $\mathbf{X} = \left[\mathbf{x}^{(1)}\ldots\mathbf{x}^{(N)}\right]$. We consider a hybrid situation where the classes of a few learning vectors are known (supervised learning), while others are of unknown class (unsupervised learning). Actually, we consider a more general case in which knowledge of $p\left(C_k/\mathbf{x}^{(n)}\right)$ could be available for a subset of $k-n$ pairs (supervised learning) and unknown for the rest of the $k-n$ pairs (unsupervised learning).

We approach the problem of estimating $\mathbf{A}_k^{-1}$, $\mathbf{b}_k$ and $p(\mathbf{s}_k)$ from a log-likelihood perspective and assuming a non-parametric model for $p(\mathbf{s}_k)$. We demonstrated that the derivates of all the unknown parameters $\mathbf{W}_k$, $\mathbf{b}_k$ ($\boldsymbol{\Psi}$ is a compact notation) can be expressed as [5],

$$\frac{\delta L(\mathbf{x}/\boldsymbol{\Psi})}{\delta \boldsymbol{\Psi}_k}=\sum_{n=1}^{N}\frac{\delta \log\left[\left|\det \mathbf{W}_k\right|\cdot p\left(\mathbf{W}_k\left(\mathbf{x}^{(n)}-\mathbf{b}_k\right)\right)\right]}{\delta \boldsymbol{\Psi}_k}\frac{\left|\det \mathbf{W}_k\right|\cdot p\left(\mathbf{W}_k\left(\mathbf{x}^{(n)}-\mathbf{b}_k\right)\right)p(C_k)}{\sum_{k'=1}^{K}\left|\det \mathbf{W}_{k'}\right|\cdot p(\mathbf{W}_k(\mathbf{x}^{(n)}-\mathbf{b}_{k'}))}$$

$$k=1\ldots K \tag{3.5}$$

Equating the derivative of the log-likelihood to zero leads to the set of equations that are to be solved in order to find the parameters $\mathbf{W}_k$, $\mathbf{b}_k$ $k=1\ldots K$. Unfortunately, a closed form solution for the set of equations obtained in this way is unattainable. First of all we need an expression for the source pdf's $p(\mathbf{s}_k)$ $k=1\ldots K$. In ICA, it is assumed that $p(\mathbf{s}_k)=p(s_{k1})\cdot p(s_{k2})\cdots p(s_{kM})$, where $\mathbf{s}_k=\left[s_{k1}\, s_{k2}\ldots s_{kM}\right]^T$ and M is the dimension of the data (number of sources). This simplifies the problem, but we still need an expression for the marginal pdf's. Once the expression for the pdf's is given, a highly nonlinear set of equations must be solved. This unavoidably leads to iterative methods, as explained in the next section.

3.2 Iterative Solutions

Iterative solutions are based on decoupling the computation of the parameters from the computation of $p(\mathbf{s}_k)$ $k=1\ldots K$ (or of any other function deduced from $p(\mathbf{s}_k)$ $k=1\ldots K$, if needed). We consider the steps given in Table 3.1.

3.3 A General Procedure for ICAMM

3.3.1 *Non-Parametric Estimation of the Source pdf's*

Non-parametric estimation of the source pdf's is the most general way of approaching this problem since no particular parametric models are needed. Imposing independence $p\left(\mathbf{s}_k^{(n)}\right)=p\left(s_{k1}^{(n)}\right)p\left(s_{k2}^{(n)}\right)\ldots p\left(s_{kM}^{(n)}\right)$, we have to estimate the marginals $p\left(s_{kM}^{(n)}\right)$ $m=1\ldots M$ $k=1\ldots K$. This can be done by means of [6]

Table 3.1 Steps for iterative solutions in ICAMM

0. Initialize $i = 0$, $\mathbf{W}_k(0)$, $\mathbf{b}_k(0)$
1. Compute $\mathbf{s}_k^{(n)}(i) = \mathbf{W}_k(i)\left(\mathbf{x}^{(n)} - \mathbf{b}_k(i)\right) \quad k = 1\ldots K \;\; n = 1\ldots N$
2. Compute $\dfrac{\delta \log p\left(\mathbf{x}^{(n)}/C_k, \boldsymbol{\Psi}_k\right)}{\delta \boldsymbol{\Psi}_k}(i) = \dfrac{\delta \log\left[\left|\det \mathbf{W}_k(i)\right| \cdot p\left(\mathbf{s}_k^{(n)}(i)\right)\right]}{\delta \boldsymbol{\Psi}_k} \quad k = 1\ldots K \;\; n = 1\ldots N.$
3. Compute $p\left(C_k/\mathbf{x}^{(n)}, \boldsymbol{\Psi}\right)(i) = \dfrac{\left|\det \mathbf{W}_k(i)\right| \cdot p\left(\mathbf{s}_k^{(n)}(i)\right) p(C_k)}{\sum_{k'=1}^{K} \left|\det \mathbf{W}_{k'}(i)\right| \cdot p\left(\mathbf{s}_{k'}^{(n)}(i)\right)} \quad k = 1\ldots K \;\; n = 1\ldots N$
4. Compute $\dfrac{\delta L(\mathbf{X}/\boldsymbol{\Psi})}{\delta \boldsymbol{\Psi}_k}(i) \quad k = 1\ldots K$ using
   $\dfrac{\delta L(\mathbf{X}/\boldsymbol{\Psi})}{\delta \boldsymbol{\Psi}_k} = \sum_{n=1}^{N} \dfrac{\delta \log p\left(\mathbf{x}^{(n)}/C_k, \boldsymbol{\Psi}_k\right)}{\delta \boldsymbol{\Psi}_k} p\left(C_k/x^{(n)}, \boldsymbol{\Psi}\right) \quad k = 1\ldots K$ and the results of steps 2 and 3
5. Actualize $\mathbf{W}_k$, $\mathbf{b}_k \quad k = 1\ldots K$ using a gradient algorithm
   $\mathbf{W}_k(i+1) = \mathbf{W}_k(i) + \alpha \dfrac{\delta L(\mathbf{x}/\boldsymbol{\Psi})}{\delta \mathbf{W}_k}(i); \quad \mathbf{b}_k(i+1) = \mathbf{b}_k(i) + \beta \dfrac{\delta L(\mathbf{x}/\boldsymbol{\Psi})}{\delta \mathbf{b}_k}(i) \quad k = 1\ldots K$
   (3.6)
   where higher values α and β increase the speed of convergence and the final error variance
6. Go back to step 1, with the new values $\mathbf{W}_k(i+1)$, $\mathbf{b}_k(i+1)$ and $i \to i+1$

$$p\left(s_{kM}^{(n)}\right) = a \cdot \sum_{n' \neq n} e^{-\frac{1}{2}\left(\frac{s_{km}^{(n)} - s_{km}^{(n')}}{h}\right)^2}, \quad m = 1\ldots M \;\; k = 1\ldots K \tag{3.7}$$

where a is a normalization constant and h is a constant that defines the degree of smoothing of the estimated pdf. Equation (3.7) must be applied at every iteration of the algorithm on the source training sets computed in step 2. Using Eq. (3.7), we can finally write [5]:

$$\frac{\delta L(\mathbf{X}/\boldsymbol{\Psi})}{\delta \mathbf{W}_k} = \sum_{n=1}^{N} \left[\left(\mathbf{W}_k^T\right)^{-1} + \mathbf{f}\left(\mathbf{s}_k^{(n)}\right)\left(\mathbf{x}^{(n)} - \mathbf{b}_k\right)^T\right] \cdot \frac{\left|\det \mathbf{W}_k\right| \cdot p\left(\mathbf{s}_k^{(n)}\right) p(C_k)}{\sum_{k'=1}^{K} \left|\det \mathbf{W}_{k'}\right| p\left(\mathbf{s}_{k'}^{(n)}\right)} \tag{3.8a}$$

$$\frac{\delta L(\mathbf{X}/\boldsymbol{\Psi})}{\delta \mathbf{b}_k} = \sum_{n=1}^{N} \left[-diag\left[\mathbf{f}\left(\mathbf{s}_k^{(n)}\right)\right] \mathbf{w}_{km}\right] \cdot \frac{\left|\det \mathbf{W}_k\right| \cdot p\left(\mathbf{s}_k^{(n)}\right) p(C_k)}{\sum_{k'=1}^{K} \left|\det \mathbf{W}_{k'}\right| p\left(\mathbf{s}_{k'}^{(n)}\right)} \tag{3.8b}$$

Another possibility is replacing the result of Eq. (3.8a) by the natural gradient. This method of optimization has demonstrated good convergence properties [7]. Thus we can write

$$\frac{\tilde{\delta}L(\mathbf{X}/\boldsymbol{\Psi})}{\tilde{\delta}\mathbf{W}_k} = \frac{\delta L(\mathbf{X}/\boldsymbol{\Psi})}{\delta \mathbf{W}_k} \cdot \mathbf{W}_k^T \cdot \mathbf{W}_k$$
$$= \sum_{n=1}^{N} \left[\mathbf{I} + \mathbf{f}\left(\mathbf{s}_k^{(n)}\right)\left(\mathbf{s}_k^{(n)}\right)^T \right] \cdot \mathbf{W}_k \cdot \frac{|\det \mathbf{W}_k| \cdot p\left(\mathbf{s}_k^{(n)}\right)p(C_k)}{\sum_{k'=1}^{K} |\det \mathbf{W}_{k'}| p\left(\mathbf{s}_{k'}^{(n)}\right)} \tag{3.8c}$$

Then we can apply Eqs. (3.8a, b, c) in the gradient updating algorithm of Eq. (3.6) to iteratively find the parameters $\mathbf{W}_k$, $\mathbf{b}_k$, $k = 1\ldots K$.

This algorithm is connected to the one proposed in [1], but here the nonlinear function $\mathbf{f}(\mathbf{s}_k^n)$ is the one corresponding to a non-parametric estimation of the source pdf. Note that both $\mathbf{f}(\mathbf{s}_k^{(n)})$ and $\mathbf{p}(\mathbf{s}_k^{(n)})$ are actually computed in a non-parametric manner (kernel density estimates) using $f\left(s_{km}^{(n)}\right) = \frac{1}{h^2}\left[\frac{\sum_{n' \neq n} s_{km}^{(n')} \cdot e^{-\frac{1}{2}\left(\frac{s_{km}^{(n)} - s_{km}^{(n')}}{h}\right)^2}}{\sum_{n' \neq n} e^{-\frac{1}{2}\left(\frac{s_{km}^{(n)} - s_{km}^{(n')}}{h}\right)^2}} - s_{km}^{(n)}\right]$ and (3.7) respectively, given a suitable algorithm to a more extensive field of applications. The estimation is asymptotically unbiased and efficient, and it is shown to converge to the true pdf under several measures, when a suitable kernel is chosen [6]. The parameter h, which controls the smoothness of the functional f, was estimated as $h = 1.06\sigma N^{-1/5}$ ($\sigma = \mathrm{std}(\mathbf{s}_m) \quad m = 1\ldots M$), which is the normal reference rule using a Gaussian kernel [6]. We use this value of h in simulations of Sect. 3.5.

3.3.2 Unsupervised–Supervised Learning

We consider a hybrid supervised–unsupervised situation in which we could know a priori $p\left(C_k/\mathbf{x}^{(n)}, \boldsymbol{\Psi}\right)$ for some $k - n$ pairs. This is the most general form to define the possible prior knowledge about the training set in a probabilistic context. For example, if $\mathbf{x}^{(n)}$ is of known class C_{k_0}, then $p\left(C_{k_0}/\mathbf{x}^{(n)}, \boldsymbol{\Psi}\right) = 1$ and $p\left(C_k/\mathbf{x}^{(n)}, \boldsymbol{\Psi}\right) = 0 \quad k \neq k_0$, but we can think of more general cases. For those $k - n$ pairs where there is prior knowledge about $p\left(C_k/\mathbf{x}^{(n)}, \boldsymbol{\Psi}\right)$, we can avoid the computation of $\frac{|\det \mathbf{W}_k| \cdot p\left(\mathbf{s}_k^{(n)}\right)}{\sum_{k'=1}^{K} |\det \mathbf{W}_{k'}| p\left(\mathbf{s}_{k'}^{(n)}\right)}$ in Eqs. (3.8a, b, c) for all the iterations, using the known $p\left(C_k/\mathbf{x}^{(n)}, \boldsymbol{\Psi}\right)$ instead.

To help in the correct initialization of the categories in this general supervised–unsupervised method, it is convenient to select the initial centroids in the form

$$\mathbf{b}_{ks}(0) = \frac{\sum_{ns} \mathbf{x}^{(ns)} p\left(C_{ks}/\mathbf{x}^{(ns)}, \boldsymbol{\Psi}\right)}{\sum_{ns} p\left(C_{ks}/\mathbf{x}^{(ns)}, \boldsymbol{\Psi}\right)} \quad ks \in [1, K] \;\; ns \in [1, N] \tag{3.9}$$

where $ks - ns$ extends only to the pairs $k - n$ where there is prior knowledge about $p\left(C_k/\mathbf{x}^{(n)}, \boldsymbol{\Psi}\right)$. For those categories to be learned in a totally unsupervised form, simply initialize the centroids randomly. Actually, Eq. (3.9) could be used in any iteration for all $k - n$ pairs as an alternative to Eq. (3.8b), as suggested in [1].

3.3.3 Using Any ICA Algorithm

The computation of the gradient in Eq. (3.8a) or (3.8c) clearly has two separate factors. The first one is related to the underlying ICA model of every class, while the second one is related to ICAMM. The first factor is $\left(\mathbf{W}_k^T\right)^{-1} + \mathbf{f}\left(\mathbf{s}_k^{(n)}\right)\mathbf{x}^{(n)T}$ or $\left[\mathbf{I} + \mathbf{f}\left(\mathbf{s}_k^{(n)}\right)\left(\mathbf{s}_k^{(n)}\right)^T\right] \cdot \mathbf{W}_k$ in Eq. (3.8a) or (3.8c) respectively. The second factor may be thought of as a weighting factor that defines the relative degree of correction of the parameters in a particular iteration: corrections are proportional to the (estimated or known) $p\left(C_k/\mathbf{x}^{(n)}, \boldsymbol{\Psi}\right)$ for every class. Recognizing this fact leads naturally to the conclusion that any of the many ICA alternative algorithms could be used.

The general ICAMM algorithm (including non-parametric source pdf estimation, supervised–unsupervised learning, and possibility of selecting a particular ICA algorithm) is in Table 3.2. The correction of residual dependence in the classification stage of the algorithm is explained in the next section.

3.3.4 Correction of the Conditioned Class-Probability After Convergence

The learning stage of the algorithm in Table 3.2 stops when iterations do not significantly change the updated parameter estimates. Let us assume that this is done at iteration $i = I$. Then, if a new feature vector is to be classified, we need to compute $p(C_k/\mathbf{x})$ using Eq. (3.4), with the final parameter and source estimates, i.e.,

$$p(C_k/\mathbf{x}) = \frac{|\det \mathbf{W}_k(I)| p(\mathbf{s}_k) p(C_k)}{\sum_{k'=1}^{K} |\det \mathbf{W}_{k'}(I)| p(\mathbf{s}_{k'}) p(C_k)} \quad \mathbf{s}_k = \mathbf{W}_k(I)(\mathbf{x} - \mathbf{b}_k(I)) \;\; k = 1, \ldots, K \tag{3.10}$$

Table 3.2 Proposed ICAMM algorithm—the Mixca procedure

Initialization

0. Initialize $i = 0$, $\mathbf{W}_k(0)$, $\mathbf{b}_k(0)$. Use Eq. (3.9) to initialize $\mathbf{b}_k(0)$ for those classes that have supervision in the training set. Select an ICA algorithm

Learning stage

1. Compute $\mathbf{s}_k^{(n)}(i) = \mathbf{W}_k(i)\left(\mathbf{x}^{(n)} - \mathbf{b}_k(i)\right) \quad k = 1\ldots K \;\; n = 1\ldots N$
2. Directly use $p\left(C_k/\mathbf{x}^{(n)}, \mathbf{\Psi}\right)(i) = p\left(C_k/\mathbf{x}^{(n)}, \mathbf{\Psi}\right)$ for those $k - n$ pairs with knowledge about $p\left(C_k/\mathbf{x}^{(n)}, \mathbf{\Psi}\right)$. Compute $p\left(C_k/\mathbf{x}^{(n)}, \mathbf{\Psi}\right)(i) = \dfrac{|\det \mathbf{W}_k(i)| \cdot p\left(\mathbf{s}_k^{(n)}(i)\right)p(C_k)}{\sum_{k'=1}^{K} |\det \mathbf{W}_{k'}(i)| p\left(\mathbf{s}_{k'}^{(n)}(i)\right)} \quad k = 1\ldots K$ for the rest of the $k - n$ pairs. Use Eq. (3.7) to estimate $p\left(\mathbf{s}_k^{(n)}(i)\right)$
3. Use the selected ICA algorithm to compute the increments $\Delta_{ICA}^{(n)} \mathbf{W}_k(i)$ corresponding to the observation $\mathbf{x}^{(n)}$, $n = 1\ldots N$, which would be applied in $\mathbf{W}_k(i)$ in an "isolated" learning of class C_k. Compute the total increment by $\Delta\mathbf{W}_k(i) = \sum_{n=1}^{N} \Delta_{ICA}^{(n)} \mathbf{W}_k(i) \cdot p\left(C_k/\mathbf{x}^{(n)}, \mathbf{\Psi}\right)(i)$. Update $\mathbf{W}_k(i+1) = \mathbf{W}_k(i) + \alpha \cdot \Delta\mathbf{W}_k(i) \quad k = 1\ldots K.$
4. Compute $\Delta\mathbf{b}_k(i)$ using $\Delta\mathbf{b}_k(i) = \sum_{n=1}^{N} \left[-diag\left[\mathbf{f}\left(\mathbf{s}_k^{(n)}\right)\right]\mathbf{w}_{km}(i) \cdot p\left(C_k/\mathbf{x}^{(n)}, \mathbf{\Psi}\right)(i)\right]$

$$\text{Use } f\left(s_{km}^{(n)}\right) = \frac{1}{h^2}\left[\frac{\sum_{n' \neq n} s_{km}^{(n')} \cdot e^{-\frac{1}{2}\left(\frac{s_{km}^{(n)} - s_{km}^{(n')}}{h}\right)^2}}{\sum_{n' \neq n} e^{-\frac{1}{2}\left(\frac{s_{km}^{(n)} - s_{km}^{(n')}}{h}\right)^2}} - s_{km}^{(n)}\right] \text{ to estimate } \mathbf{f}\left(\mathbf{s}_k^{(n)}\right)$$

Actualize $\mathbf{b}_k(i+1) = \mathbf{b}_k(i) + \beta \cdot \Delta\mathbf{b}_k(i) \; k = 1\ldots k$, or re-estimate

$$\mathbf{b}_k(i+1) = \frac{\sum_{n=1}^{N} \mathbf{x}^{(n)} p\left(C_k/\mathbf{x}^{(n)}, \mathbf{\Psi}\right)(i)}{\sum_{n=1}^{N} p\left(C_k/\mathbf{x}^{(n)}, \mathbf{\Psi}\right)(i)} \quad k = 1\ldots K$$

5. Go back to step 1, with the new values $\mathbf{W}_k(i+1)$, $\mathbf{b}_k(i+1)$ and $i \to i+1$

Classification stage

6. Assuming learning stage stops at iteration $i \to I$. For a new feature vector to be classified, estimate $p(C_k/\mathbf{x}) = \dfrac{|\det \mathbf{W}_k(I)| p(\mathbf{s}_k) p(C_k)}{\sum_{k'=1}^{K} |\det \mathbf{W}_{k'}(I)| p(\mathbf{s}_{k'}) p(C_k)} \quad \mathbf{s}_k = \mathbf{W}_k(I)(\mathbf{x} - \mathbf{b}_k(I)) \;\; k = 1, \ldots, K$
7. Estimate the source pdf using

$$p(\mathbf{s}_k) = p(s_{k1}) \cdot p(s_{k2}) \cdots p(s_{kM}) \text{ where } p(s_{km}) = a \cdot \sum_{n=1}^{N} e^{-\frac{1}{2}\left(\frac{s_{km} - s_{km}^{(n)}(I)}{h}\right)^2} \text{ or using}$$

multidimensional density estimation (considering residual dependence)

$$p(\mathbf{s}_k) = a_0 \cdot \sum_{n=1}^{N} e^{-\frac{1}{2}\left(\frac{\left[\mathbf{s}_k - \mathbf{s}_k^{(n)}(I)\right]^T \left[\mathbf{s}_k - \mathbf{s}_k^{(n)}(I)\right]}{h_0^2}\right)}$$

Table 3.3 A notation for different variants of the Mixca procedure

ICAMM parameter updating algorithm	Mixca variant notation
Non-parametric estimation	Mixca, non-parametric Mixca
JADE	Mixca-JADE
FastIca	Mixca-FastIca
TDSEP	Mixca-TDSEP
InfoMax	Mixca-InfoMax

In principle, the source pdf in Eq. (3.10) can also be estimated in a non-parametric manner following

$$p(\mathbf{s}_k) = p(s_{k1}) \cdot p(s_{k2}) \cdots p(s_{kM}) \quad \text{where } p(s_{km}) = a \cdot \sum_{n=1}^{N} e^{-\frac{1}{2}\left(\frac{s_{km} - s_{km}^{(n)}(I)}{h}\right)^2} \tag{3.11}$$

However, although independence is imposed by means of the underlying ICA algorithm that is selected to learn the mixtures, there is no guarantee that the final source vectors have independent elements. Some residual dependence may still remain due to possible nonlinearities that are not accounted for in the basic model in Eq. (3.1). If this is the case, Eq. (3.11) will produce erroneous estimates of the source pdf, and thus incorrect computing of the probability of every class conditioned to the feature vector (posterior probability) by applying Eq. (3.10). This suggests the convenience of improving the estimation of the source pdf by considering general methods of multidimensional density estimation, preferably of the non-parametric type [8], in order to keep the generality of the proposed framework. A rather simple possibility is to consider the estimator

$$p(\mathbf{s}_k) = a_0 \cdot \sum_{n=1}^{N} e^{-\frac{1}{2}\left(\frac{\left[\mathbf{s}_k - \mathbf{s}_k^{(n)}(I)\right]^T \left[\mathbf{s}_k - \mathbf{s}_k^{(n)}(I)\right]}{h_0^2}\right)} \tag{3.12}$$

If this post-convergence correction is applied, the analyzers could no longer be independent. Therefore, we have developed a procedure that analyzes mixture data with components that may or may not be independent, which we call Mixca (Mixture of Component Analyzers). This procedure allows all the parameters involved in ICAMM (mixture matrices, centroids, and source probability densities) to be estimated. The Mixca procedure defines a framework with a set of variants depending on the embedded algorithm used for ICAMM parameter updating. A notation for different variants of the Mixca procedure is in Table 3.3.

3.3.5 Discussion

At this point, we would like to point out that the algorithm above is intended to extend the ICAMM to a general framework. This is done by including new enhancements in ICAMM such as semi-supervision, correction of residual dependencies, embedding of any ICA algorithm in the learning process, and non-parametric estimation of the source densities. Let us discuss on some of the features of the proposed algorithm.

We have selected a kernel non-parametric estimator to estimate the pdf (see Eq. (3.7)). The kernel estimator has a closed form expression so that subsequent analytical development is possible, as we have done in [5]. This is a well-studied and simple method which has the advantage that the estimated density can be easily guaranteed to be a true density, i.e., it is nonnegative and integrates to 1 [8]. Obviously, other alternatives having closed forms [9] would be readily applicable in ICAMM by simply changing the corresponding expressions $p\left(\mathbf{s}_k^{(n)}\right)$, $\frac{\delta \log p\left(\mathbf{s}_k^{(n)}\right)}{\delta \mathbf{W}_k}$, and $\frac{\delta \log p\left(\mathbf{s}_k^{(n)}\right)}{\delta \mathbf{W}_k}$. A comparison among different pdf estimators and their influence in ICAMM is outside the scope of this work. Actually, there are not yet many works devoted to comparing how different pdf non-parametric estimators influence the performance of ICA.

We have selected a gradient algorithm for optimization to iteratively search the maximum likelihood solution. Gradient algorithms are simple to implement and have reasonably good convergence properties, particularly in combination with ad hoc techniques to avoid blocking in local minima. To this end, we used an annealing method in the implementation of the algorithm. The stepsize or learning rate was annealing during the adaptation process in order to provide faster and proper convergence. In addition, the learning rule of Eq. (3.8c) was used in the algorithm implementation in order to take advantage of the efficiency in learning of the natural gradient technique. The natural gradient is based on differential geometry and employs knowledge of the Riemannian structure of the parameter space to adjust the gradient search direction. Furthermore, natural gradient is asymptotically Fisher-efficient for maximum likelihood estimation [7].

Alternatives to gradient algorithms are possible in ICAMM, but we think it is more interesting to understand to what extent the different convergence analyses and experiments previously considered in ICA [10, 11] generalize to ICAMM. Note that, as indicated in step 3 of the iterative algorithm in Sect. 3.3.3, the updating increment of $\mathbf{W}_k$ in every iteration is a weighted sum of the separate increments due to every training sample vector. The corresponding weights are the computed probability of the training vector belonging to class k. We can write the increment in every iteration in the form

$$\begin{aligned}
\Delta \mathbf{W}_k(i) &= \sum_m \Delta_{ICA}^{(n)} \mathbf{W}_k(i) \cdot p\left(C_k/\mathbf{x}^{(n)}, \mathbf{\Psi}\right)(i) + \sum_l \Delta_{ICA}^{(n)} \mathbf{W}_k(i) \cdot p\left(C_k/\mathbf{x}^{(n)}, \mathbf{\Psi}\right)(i) \\
&= \sum_m \Delta_{ICA}^{(m)} \mathbf{W}_k(i) + \sum_m \Delta_{ICA}^{(n)} \mathbf{W}_k(i) \cdot \left(1 - p\left(C_k/\mathbf{x}^{(m)}, \mathbf{\Psi}\right)(i)\right) + \sum_l \Delta_{ICA}^{(l)} \mathbf{W}_k(i) \cdot p\left(C_k/\mathbf{x}^{(l)}, \mathbf{\Psi}\right)(i) \\
&= \left[\sum_m \Delta_{ICA}^{(m)} \mathbf{W}_k(i)\right] + r_k(i)
\end{aligned} \tag{3.13}$$

where m and l are indices that correspond to the training vectors belonging to class k and training vectors not belonging to class k, respectively.

In a totally supervised training $p(C_k/\mathbf{x}^{(m)}, \boldsymbol{\Psi})(i) = 1\ \forall m$ and $p(C_k/\mathbf{x}^{(l)}, \boldsymbol{\Psi})(i) = 1\ \forall l$; hence the residual increment $r_k(i) = 0\ \forall k$. In other words, only the training vectors corresponding to class k will affect the increment, and the convergence properties will be the same as the properties of the underlying ICA algorithm: every ICA model of the mixture is separately learned. In the case of semi-supervised learning, the lower the amount of supervision, the higher the perturbation introduced by $r_k(i)$ in the learning of class k. It is clear that close to the optimum, $p(C_k/\mathbf{x}^{(m)}, \boldsymbol{\Psi})(i) \simeq 1\ \forall m$ and $p(C_k/\mathbf{x}^{(l)}, \boldsymbol{\Psi})(i) \simeq 1\ \forall l$; that is, the convergence properties will essentially be the same as those of the underlying ICA algorithm. Since it is very difficult to undertake a general convergence analysis of what happens far from the optimum in ICAMM, we have included some simulations in the next section. These simulations demonstrate how the percentage of semi-supervision modifies the learning curves that correspond to the totally supervised case and how this percentage modifies the approximate number of observation vectors in order to achieve a particular mean SIR (signal-to-interference ratio), which is defined in next section.

Finally, we have defined the demixing matrix for every class $\mathbf{W}_k = \mathbf{A}_k^{-1}$, $(k = 1 \ldots K)$, that is, a prewhitening step is not included. In several ICA algorithms, the data are first whitened as explained in Chap. 2. Prewhitening is an optional step that has been used in many ICA algorithms allowing computational-burden reduction [11, 12]. However, also there exists algorithms that avoid the prewhitening phase, see for instance [13, 14]. In the context of semi-supervised ICAMM, it is not possible to prewhiten the data because a priori the allocation of the points to each class k is not totally known [15]. Specifically, in our Mixca procedure for unsupervised or semi-supervised learning, the estimation of the whitening matrix can be incorporated in each of the parameter updating steps. In the case of supervised learning, the prewhitening step can be performed previous to the Mixca procedure. We performed several simulations with Laplacian and Uniform distributed sources, which demonstrated that there are no major differences in BSS (SIR $\leq$ 1 dB) whether or not a prewhitening step is used.

3.4 Simulations

In this section, we provide a demonstration of the performance of the algorithms proposed in previous sections. Several simulations with different kinds and numbers of source densities and numbers of classes (ICA mixtures) are presented for the evaluation of the proposed technique in BSS, classification of ICA mixtures, convergence properties, classification of ICA mixtures with nonlinear dependencies, and semi-supervised learning. The parameters of Mixca were configured as follows: $\alpha = 5\mathrm{e}-05$ (Eq. 3.6); $\alpha = 1$ (Eq. 3.7); $h = 1.06\sigma N^{-1/5}(\sigma = \mathrm{std}(\mathbf{s}_m)$

$m = 1\ldots M$)(Eq. 3.7); likelihood threshold $= (L(i) - L(i-1))/L(i) = 1\mathrm{e}-05$ (i = number of the current iteration and L is estimated using equation $L(\mathbf{X}/\mathbf{\Psi}) = \log p(\mathbf{X}/\mathbf{\Psi}) = \sum_{n=1}^{N} \log p(\mathbf{x}^{(n)}/\mathbf{\Psi})$).

3.4.1 Performance in BSS

The ICA mixture model is a kind of nonlinear ICA technique that extends the linear ICA method by learning multiple ICA models and weighting them in a probabilistic manner [1]. Approaching blind source separation of ICA mixture data using standard ICA algorithms could give an inaccurate estimation of the sources. In this section, we investigate the performance in blind source separation of the non-parametric Mixca and the following ICA algorithms: FastIca [16], JADE [17], TDSEP [18], InfoMax [19], Extended InfoMax [14], Kernel-ICA [20], Npica [21], Radical [22]. Two simulation experiments of BSS were attempted with ICA mixture and ICA generative data models. The separation efficiency was estimated by the mean SIR, defined as $10\log_{10}\left(\sum_{n=1}^{N} s^{(n)^2} \Big/ \sum_{n=1}^{N} (\hat{s}^{(n)} - s^{(n)})^2\right)$ (dB), where $s^{(n)}$ is the original source signal and $\hat{s}^{(n)}$ is the reconstructed source signal.

The first experiment consisted of two ICA mixtures, each of which had three components (M=3) composed by Laplacian with a sharp peak at the bias and heavy tails and/or uniform distributed sources generated randomly. The observation vectors were obtained adding together the data generated from the sources of the two ICAs, i.e., the first part of the observation vectors corresponded to data generated by the first ICA, and the second part of the observation vectors were the data generated by the second ICA. From the observation vectors, the ICA algorithms estimated 3 sources for one theoretical underlying ICA whereas the Mixca algorithm, configured with K=2, estimated two sets of 3 sources for two ICAs. A total of 300 Montecarlo simulations of the first experiment were performed for each of the following values of N = 200, 500, 1000, 2000. The mean results for the first simulation experiment are shown in Fig. 3.1. The Mixca algorithm that assigns the observation vectors to different sources of two ICA models outperforms the standard ICA algorithms, which attempt to estimate a single set of sources for one ICA model for all the observation vectors of the mixture. The evolution curves of SIR values for the ICA algorithms show no improvement with the increase in the number of observation vectors for pdf estimation, and the SIR levels below 8-10 dB thresholds are indicative of a failure to obtaining the desired source separation. In contrast, the match between estimated and original sources increases significantly with a higher number of observation vectors used in pdf estimation for the Mixca algorithm.

The second experiment consisted of one-ICA datasets with mixtures of the five different sources shown in Table 3.4. The SIR was used to measure algorithm

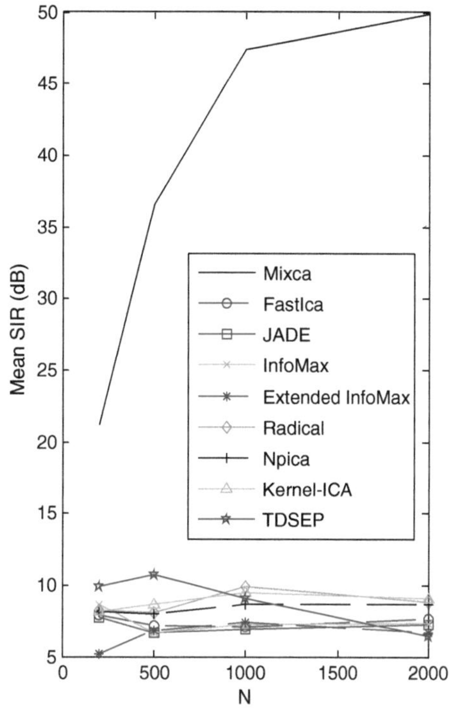

Fig. 3.1 First simulation experiment—BSS in ICA mixture datasets

Table 3.4 Sources used in the simulation experiment

Source #	Source type	Skewness	Kurtosis	*pdf plot*
1	Uniform	0.0	−1.2	
2	Laplacian ($b = 1$)	0.0	3	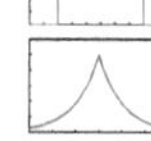
3	Normal	0.0	0.0	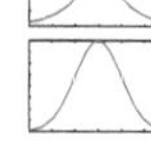
4	Rayleigh ($\beta = 1$)	0.631	0.245	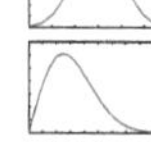
5	K-noise ($m = 1$)	1.28	5.38	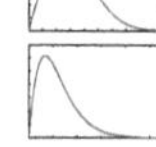

performance in BSS, and the Mixca algorithm was configured to estimate the parameters for only one ICA. TDSEP, which is an algorithm based on exploiting signal time structure, was not tested in the second experiment since this experiment included i.i.d. sources. If the sources involved in the problem have a time structure, the second order based methods as TDSEP are appropriate. In this case, the limitation is not due to the Gaussianity of the signals; it is due to the correlation functions. These functions can extract Gaussian signals, but with different spectra by diagonalizing correlation matrices for different time lags. Second order based methods are different from ICA traditional methods, which recover non Gaussian

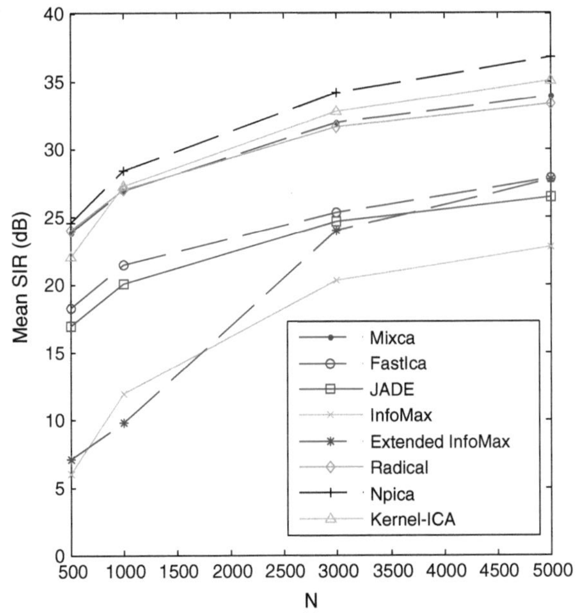

Fig. 3.2 Second simulation experiment—BSS in ICA datasets

random variables (at most one Gaussian) by approximating their independence with a distance measure (contrast) obtained with higher order statistics.

A total of 300 Monte Carlo simulations of the second experiment were performed for each of the values of $N = 500, 1000, 3000, 5000$. The mean results of the second experiment are shown in Fig. 3.2.

The mean results for the second experiment (Fig. 3.2) show that non-parametric ICA algorithms outperform more conventional ICA methods. The average performance difference between the first three non-parametric algorithms is below 1.8 dB. The Mixca algorithm conducts a consistent separation improvement for different observation vector sizes, and it is able to learn the source densities for even a small number of observation vectors (500). JADE and FastIca deliver a similar performance loss with non-parametric Mixca of over 6.5 dB for all the different observation vector sizes. Extended InfoMax and InfoMax have worse separation with smaller observation vector sizes ($N = 500$, 1000 observation vectors). They achieve a performance loss over 16.9 and 16.4 dB, with respect to non-parametric Mixca. For larger observation vector sizes ($N = 3000$, 5000 observation vectors), the performance loss with non-parametric Mixca is over 7 and 11.4 dB. Higher order statistics methods exploit the non Gaussianity of the sources in order to obtain the solution. However, they do not consider the time structure of the signals, i.e., the sources are considered random variables, and not time series.

Figure 3.3 shows detailed results of the separation for each of the sources in Table 3.4. In general, non-parametric ICA algorithms consistently outperform the standard ICA algorithms for all the sources with higher gain for skewed sources

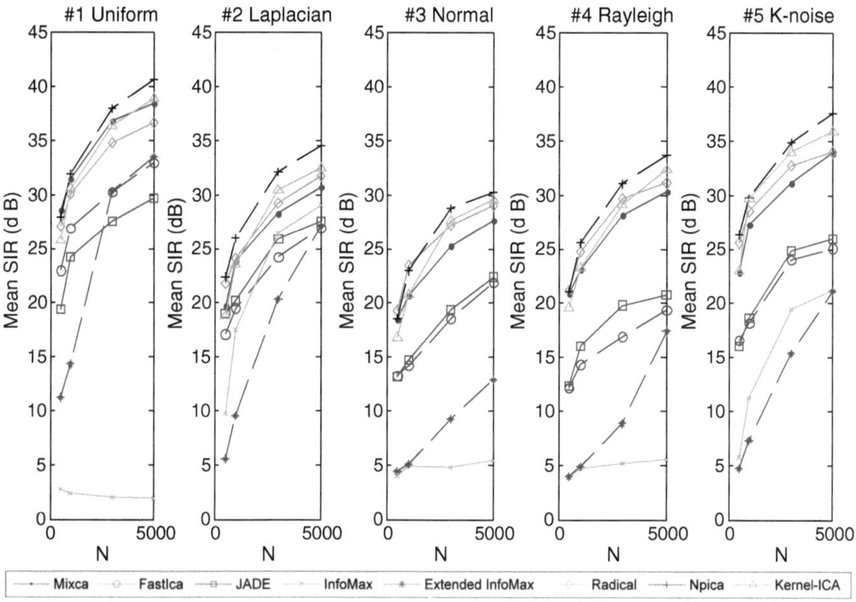

Fig. 3.3 Second simulation experiment—accuracy in BSS for the sources of Table 3.4

(Sources #4 and #5) and Gaussian sources (Source #3). For standard ICA algorithms, JADE and FastIca are more stable for all the distributions, whereas InfoMax and Extended InfoMax perform well, but with larger observation vector sizes. For Laplacian sources (Source #2), all the algorithms show a similar performance.

3.4.2 Classification of ICA Mixtures

The proposed procedure Mixca was tested with several ICA mixture datasets varying the following parameters: (i) Supervision ratio = 0,0.1,0.3,0.5,0.7,1 (number of training observation vectors/N). (ii) ICA algorithm for updating model parameters = JADE, non-parametric Mixca. (iii) Number of ICA mixtures = 2,3,4 with $N = 500$, (iv) Number of components = 2,3,4. The classes were generated randomly by mixtures of uniformly and Laplacian distributed sources with a sharp peak at the bias and heavy tails, and 400 observation vectors were used for pdf estimation. The parameters were randomly initialized, and the algorithm normally converged after 100–150 iterations depending on the initial conditions. The learning mixture algorithm was trained using 30 % of the data, obtaining the parameters $\mathbf{W}_k$ and $\mathbf{b}_k$.

Classification was performed estimating the posterior probabilities $p(C_k/\mathbf{x})$ for each testing observation vector using Eq. (3.10) and applying the learned parameters and the non-parametric source pdf estimator of Eq. (3.11). The class of

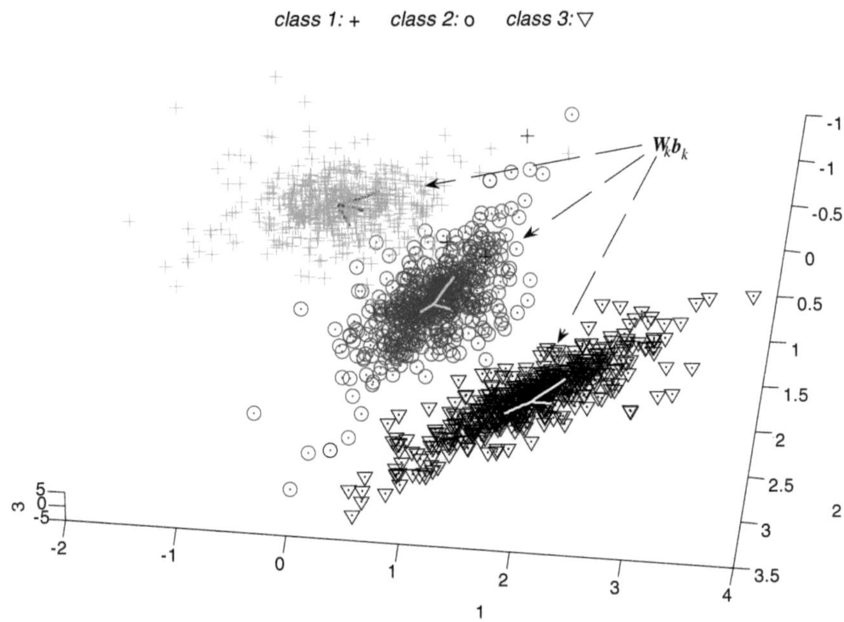

Fig. 3.4 Three ICA mixtures in a three-component space

highest probability was assigned to each observation vector. Figure 3.4 shows the classification results for one of the simulated cases with 0.1 of supervision ratio and non-parametric Mixca for updating parameters. Figure 3.4 shows three classes with mixtures of three Laplacian pdf's. The estimated parameters (basis vectors and bias terms) for each class are depicted on the data. In this case, the procedure was able to find the right direction and location of the parameters. The percentage of success in classification was 98.3.

3.4.3 Convergence Properties

The convergence properties of the procedure were empirically tested in two experiments in order to evaluate the effects of the residual increment $r_k(i) \quad k = 1 \ldots K$ of Eq. (3.13) in different semi-supervised cases. In the first experiment, we evaluated the log-likelihood, the SIR, and the classification accuracy achieved by the procedure. A total of 1000 Monte Carlo experiments were performed using the following parameters: (i) Number of classes in the ICA mixture $K = 3$; (ii) Number of observation vectors per class $N = 400$ (70 % used for training); (iii) Number of sources = 2 (Laplacians with a sharp peak at the bias and heavy tails); (iv) Supervision ratio (sr) = 0, 0.1, 0.2, 0.3, 0.4, 0.5, 0.6, 0.7, 1; (v) Embedded ICA algorithm = Non-parametric Mixca.

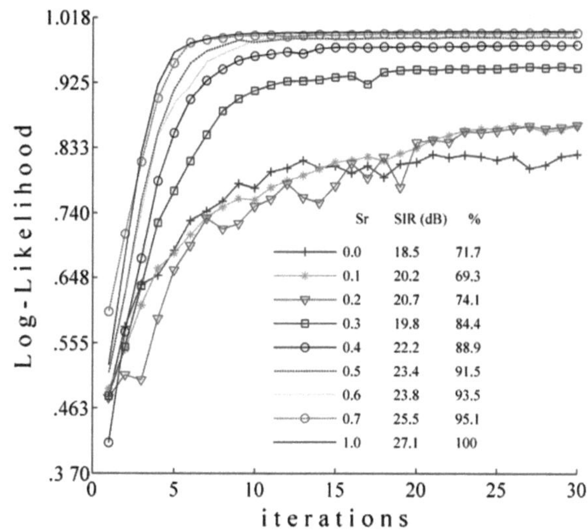

Fig. 3.5 Data log-likelihood evolution through non-parametric Mixca iterations. *Sr* supervision ratio, *SIR* signal-to-interference ratio, % classification accuracy

Figure 3.5 shows the results of the first experiment. Different graphs of the mean data log-likelihood evolution through the Mixca procedure iterations that correspond to the supervision ratios used in the training stage are depicted. The curves present an ascending behaviour with an increasing number of iterations; the adjusted parameters fit the observed data and the log-likelihood is maximized. The convergence of the algorithm depended on the supervision ratio, the overlapping areas of the classes, and the parameter initialization. The algorithm converged successfully in 99.7 % of the cases in less than 30 iterations obtaining different values of log-likelihood depending on the supervision ratio (values of log-likelihood are normalized in Fig. 3.5). The higher the values of supervision, the higher the values of log-likelihood obtained. The non-convergence cases (when the algorithm became stuck in a local maximum) corresponded to the lowest supervisions (0, 0.1); after labelling some of the data, the algorithm converged to the correct solution. In real application contexts, when there are enough historical data, this will not be a problem.

The results of Fig. 3.5 can be grouped into two sets: middle- and high-log-likelihood results. The results of log-likelihood were consistent with the results of BSS and classification. Thus, the higher the log-likelihood value, the higher SIR and the higher the classification accuracy. The graphs corresponding to the lowest sr (0, 0.1, 0.2) showed oscillating patterns, while the graphs of the highest sr (≥ 0.3) were smooth and increased monotonically with a few exception points. Thus, the uncertainty of the convergence was higher for the lowest supervisions. In addition, convergence velocity was higher for the highest supervisions, approaching the global maximum faster. In the case of the lowest supervisions, the algorithm slowly improved the data log-likelihood. Even though the algorithm surpassed zones of several local maxima, it did not converge accurately to the global maximum. Natural gradient with stepsize annealing improves the data log-likelihood providing a good compromise between convergence velocity and computational payload. The final results of convergence were consistent for all the supervision ratios and the log-likelihood results were ordered by sr. In addition, we verified that the distances (measured by MSE) between the estimated centroids $\mathbf{b}_k\ k = 1\ldots3$ and the original centroids of the ICA mixtures decrease with higher supervision.

The above results demonstrate that the perturbation introduced by $r_k(i) \quad k = 1\ldots K$ due to unlabelled data affects the convergence properties in the learning of the class parameters. This residual increment affects the cases with the lowest supervisions the most. For the highest supervisions, the convergence depends on the algorithm used to update the ICA parameters of the classes, as discussed in Sect. 3.3.5.

The classification and BSS results for sr ≥ 3 achieved the correct solution, and the results for sr < 3 were close to the correct solution. The maximum difference for different sr was the difference between the unsupervised case and the supervised case (0.176 log-likelihood, 8.6 dB SIR, 28.3 % classification accuracy). These parameters can be relatively critical depending on the specific application, and they underscore the importance of incorporating semi-supervised learning in ICAMM in order to take advantage of a partial labelling of the data (see Sect. 3.4.5). In addition, we repeated this experiment, but we changed the embedded ICA algorithm using standard algorithms such as JADE and FastIca for parameter updating. In general, the results of SIR, and classification accuracy were comparable for all the embedded algorithms. The efficiency of JADE and FastIca in separation of super-gaussian sources is known; for this kind of sources, the kernel density estimation obtained similar results. However, for the log-likelihood results, the non-parametric Mixca converged to highest values in a range of sr (0.3–1) while Mixca-JADE and Mixca-FastIca only converged to the highest values of log-likelihood for the supervised case. Thus, for these latter algorithms, more cases of middle-log-likelihood were obtained.

In the second experiment we measured the approximate number of observation vectors required by the Mixca procedure to achieve particular mean SIRs. A total of 400 Monte Carlo simulations were generated with the following parameters: (i) Number of classes in the ICA mixture $K = 2$; (ii) Number of observation vectors per class $N = 100, 200, 300, 400, 500$; (iii) Number of sources = 4 (Laplacians with a sharp peak at the bias and heavy tails); (iv) Supervision ratio (sr) = 0, 0.1, 0.3, 0.5, 0.7, 1; (v) Embedded ICA algorithm = Non-parametric Mixca.

Figure 3.6a, b show the detailed and mean results of the second experiment. Different graphs of the SIR obtained for different numbers of observation vectors that correspond to the supervision ratios used in the training stage are depicted in Fig. 3.6b. The number of observation vectors required to obtain a particular SIR value increased with less supervision. In general, the results demonstrate that the non-parametric Mixca procedure is able to achieve a good quality of SIR with a small number of observation vectors, e.g., 20 dB of SIR were obtained with only 203, 215, 231, 258, 306 observation vectors for sr = 1, 0.7, 0.5, 0.3, 0.1, respectively. The results of this experiment confirm that the convergence efficiency of the proposed procedure increases significantly when only a small

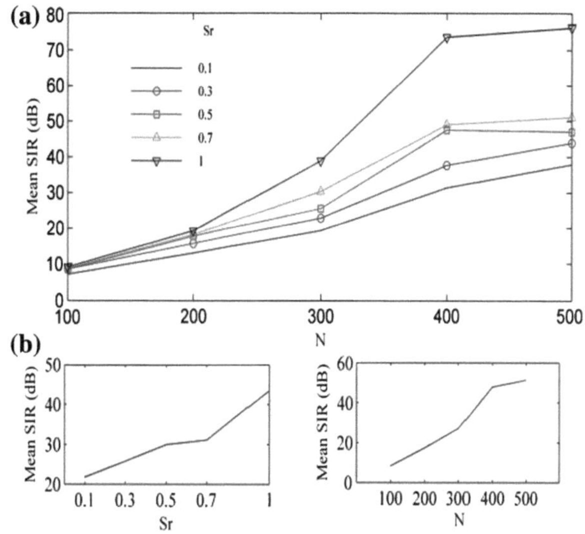

Fig. 3.6 SIR versus number of observation vectors used for density estimation. *Sr* supervision ratio

percentage of supervision is added. For instance, in the case of 500 observation vectors the SIR gain is 9.16 dB from sr = 0 to sr = 0.5 and 25.04 dB from sr = 0.7 to sr = 1.

3.4.4 Classification of ICA Mixtures with Nonlinear Dependencies

Several simulations were performed using ICA mixtures with nonlinear data. We used multi-layer perceptron (MLP) networks to model the nonlinear mixing mapping $\mathbf{g}(\cdot)$ from source signals to observations, $\mathbf{x} = \mathbf{g}(\mathbf{s})$, where $\mathbf{X}$ is the observed m-dimensional data vector, $\mathbf{g}$ is a m-component mixing function, and $\mathbf{s}$ is a n-vector of independent components. MLP networks have the universal approximation property [23] for smooth continuous mappings [24–26].

ICA mixture data were generated for two classes from uniform sources. The generative data model for one of the classes included only linear dependencies, while the other class was modelled using nonlinearities. For this latter class, the data were generated through a nonlinear mapping which was obtained by using a randomly initialized MLP network having 20 hidden neurons with the output neurons being equal to the number of sources. A total of 800 Monte Carlo experiments were performed with the following parameters: (i) Number of classes in the ICA mixture $K = 2$. (ii) Number of observation vectors per class $N = 400$. (iii) Number of sources = 2,3,4,5. (iv) Supervision ratio = 0.5. (v) Number of training observation vectors = $0.7N$.

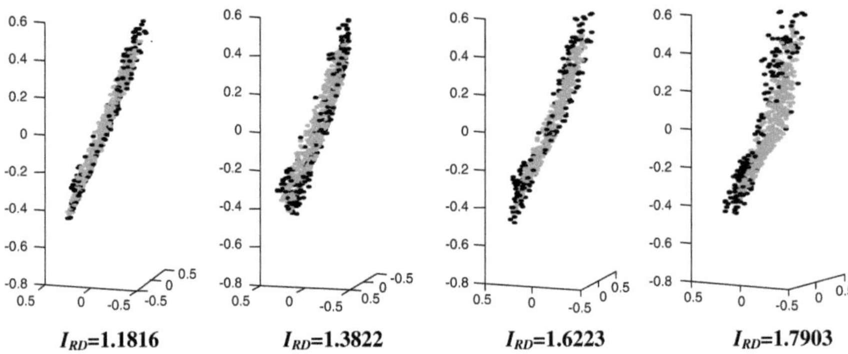

Fig. 3.7 A class with three sources altered by different grades of residual dependencies

Monte Carlo datasets were prepared, adapting the MLP to yield different grades of nonlinearities in the data. Thus, depending on the nonlinearity mapping, there were data with different degrees of adjustment to the source pdf's of the ICA model. In order to test the correction of residual dependencies, 30 % of the observation vectors with the lower values of source pdf, $p(\mathbf{s}_k^{(n)}) \quad k = 1 \ldots K\ n = 1 \ldots N$, were selected for the testing stage. The remaining 70 % of the observation vectors were selected for the training stage. The parameters of the ICA mixture were estimated using the data that best fit into the ICA model, and the classification was made on the data that supposed were to be distorted with residual dependencies. We denote any training observation vector as $\mathbf{x}_{training}$, and we denote any testing observation vector as $\mathbf{x}_{testing}$, with $p(C_k/\mathbf{x}_{training})$ and $p(C_k/\mathbf{x}_{testing})$ being their respective posterior probabilities. An index that measures the residual dependencies (I_{RD}) can be defined as the ratio between the mean of the posterior probability estimated in training ($\bar{p}(C_k/\mathbf{x}_{training}) \quad k = 1 \ldots K$) and the mean of the posterior probability estimated in testing ($\bar{p}(C_k/\mathbf{x}_{testing}) \quad k = 1 \ldots K$),

$$I_{RD} = \frac{p(C_k/\mathbf{x}_{training})}{p(C_k/\mathbf{x}_{testing})} \quad k = 1 \ldots K \tag{3.14}$$

I_{RD} tends to 1 in the case of low or null nonlinearities, i.e., all the data are well-adjusted to the ICA model. Conversely, if the data for testing do not fit the linear ICA model, I_{RD} increases since the posterior probability for testing data decreases.

Figure 3.7 shows some examples of generated datasets altered by nonlinearities for a class in a dimensional space defined by three sources. The points in grey represent the linear part of the data, and the points in black represent the points that are most altered by nonlinearities. Note the different degrees of nonlinearities and the corresponding index of residual dependencies I_{RD} estimated after classification for each of the graphs in Fig. 3.7.

The results are summarized in Fig. 3.8. The improvement in classification by the correction is presented through different values of the residual dependency index and for different numbers of sources ($D = 2, \ldots, 5$). Multi-dimensional

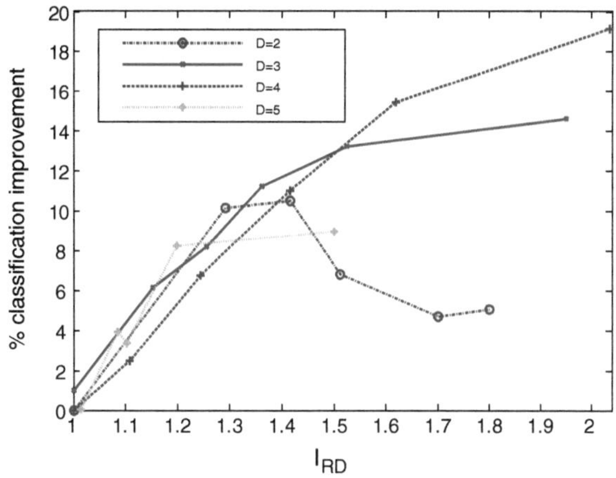

Fig. 3.8 Correction of the posterior probability in classification for different dimensions and residual dependencies

estimation of the source pdf's performed well for three and four dimensions. The higher the values of residual dependencies, the greater the improvement in classification by the correction of the posterior probability. In addition, the improvement was better in the higher four-dimensional space since one- and multi-dimensional estimations were more differentiated when the number of dimensions was increased. For two dimensions, the behaviour of the correction was not consistent for different ranges of residual dependencies, performing worse for higher values [1.52–1.8] of I_{RD} than for lower values [1.29–1.42] of I_{RD}. However, the correction always improves the classification results in all the explored ranges of residual dependencies. The varying results in the two-dimensional space were due to similar achievements by the one-dimensional and multi-dimensional estimators for lower dimensions. In the case of five and higher dimensions, a higher number of observation vectors was required for a precise non-parametric estimation of the data distribution. The results for $D = 5$ were obtained by increasing the number of observation vectors per class from 200 to 1000. In this case, the tendency of the curve was correct, but the improvement in classification was lower than for $D = 3$ and $D = 4$. A much higher number of observation vectors was required to obtain better results than those used in lower dimensions.

3.4.5 Semi-supervised Learning

The proposed procedure includes learning with both labelled and unlabelled data, which is called semi-supervised learning. Statistical intuition advises that it is reasonable to expect an average improvement in classification performance for any increase in the number of samples (labelled or unlabelled). There is applied work on semi-supervised learning that optimistically suggests the use of unlabelled data whenever available [27–30]. However, the importance of using correct modelling assumptions when learning with unlabelled data has been emphasized for Bayesian network classifiers. When the assumed probabilistic model does not match the true data generating distribution, using unlabelled data can be detrimental to the classification accuracy [31]. In addition, there is interest in studying the data that do not match underlying generative models as outlier data, whether those data should be retained (represent phenomena of interest) or rejected (they are mistakes) [32].

In this section, we explore the behaviour of the proposed ICA mixture-based classifier in both the training stage and the classification stage, depending on which data are labelled. Thus, two kinds of data were discerned for labelling the data that better fit into the ICA model and the data that did not adapt as well into the model. The latter data are considered as a kind of outlier. In addition, ICA mixtures were divided into two groups, depending on the strength of the membership of the data to the classes, i.e., the values of posterior probability $p(C_k/\mathbf{x})$. These groups were called high-fuzziness and low-fuzziness ICA mixtures, applying a term used in fuzzy classification literature [33].

A total of 200 Monte Carlo simulations were performed to generate different ICA mixture datasets with three classes, 400 observation vectors per class, two Laplacian distributed sources with a sharp peak at the bias and heavy tails; and 200 observation vectors were used for pdf estimation. For each dataset, 16 cases were obtained varying the following parameters for the training data: (i) Supervision ratio = 0.1,0.3,0.5,0.7. (ii) ratio of labelled outliers = 0,1/3,2/3,1 (number of labelled outlier observation vectors/total number of outlier observation vectors). The outlier observation vectors were mixed with Type K noise ($m = 3$) [34] to highlight their difference with the ICA data model. The data were divided as follows: 70 % for training and 30 % for testing. The parameters $\mathbf{W}_k$ and $\mathbf{b}_k$ of the generative data model for a dataset, the corresponding $\mathbf{s}_k$, and $p(C_k/\mathbf{x})$ were used as reference to estimate the accuracy in classification and the lack of adjustment compared to the corresponding parameters obtained for the 16 cases of semi-supervised training for that dataset.

The fuzziness F for a dataset was calculated as the unity minus the mean of the maxima of the posterior probability for each class,

$$F = 1 - \overline{\max}(p(C_k/\mathbf{x})) \quad k = 1 \ldots K \tag{3.15}$$

Values of F range from 0 (no fuzziness data mixture) to $1 - 1/K$ (completely fuzziness data mixture). When $F = 0$, for every observation vector the posterior probability $p(C_k/\mathbf{x})$ is 1 for a class and 0 for the other classes. When $F = 1 - 1/K$, the posterior probabilities are equally-probable for every class and observation vector. The lack of adjustment between the reference model for a dataset and a case of semi-supervised training for that dataset was measured using the Kullback–Leibler (KL) divergence D_{KL} [35],

$$D_{KL} = \int q(\mathbf{S}, \psi/\mathbf{X}) \log \frac{q(\mathbf{S}, \psi/\mathbf{X})}{p(\mathbf{S}, \psi/\mathbf{X})} d\psi d\mathbf{S} \tag{3.16}$$

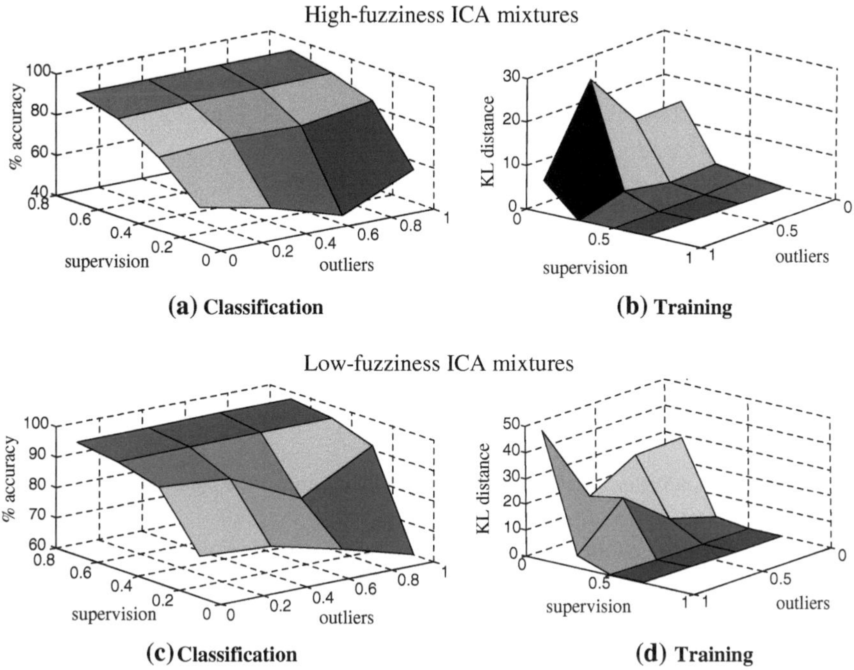

Fig. 3.9 Semi-supervised results in classification and training. **a**, **c** Classification; **b**, **d** training

where **X** is the set of available mixture data and **S** the respective source vectors. ψ denote all the unknown parameters of the mixture data model; $p(\mathbf{S}, \psi/\mathbf{X})$ denote the posterior pdf for the reference model, and $q(\mathbf{S}, \psi/\mathbf{X})$ denote the posterior pdf for the estimated model.

The mean results of the performance of semi-supervised learning are shown in Fig. 3.9 for the simulated datasets divided into high-fuzziness and low-fuzziness ICA mixtures. Since the simulations consisted of ICA mixtures of three classes, the highest value of fuzziness possible was 1–1/3, so we set 0.14 (21 % of the highest fuzziness) as the threshold between high and low fuzziness.

The evolution of the values of classification success in Fig. 3.9a shows that small increments in supervision rapidly increase the percentage of classification accuracy. They also increase the similarity, which is measured by the KL-distance between the reference models and the semi-supervised training models as shown in Fig. 3.9b. The results for low-fuzziness ICA mixtures (Fig. 3.9c, d) show a behaviour similar to that for high-fuzziness ICA mixtures. However, classification accuracy is improved with smaller increments of supervision, and the effect of outliers in worsening semi-supervised training models is higher. The data of low-fuzziness mixtures are more separated from each others, and they have a high probability of belonging to different classes. Therefore, clusters of low-fuzziness mixtures are better defined than data of high-fuzziness mixtures. Due to the low values of posterior probability in the data of high-fuzziness ICA mixtures, the

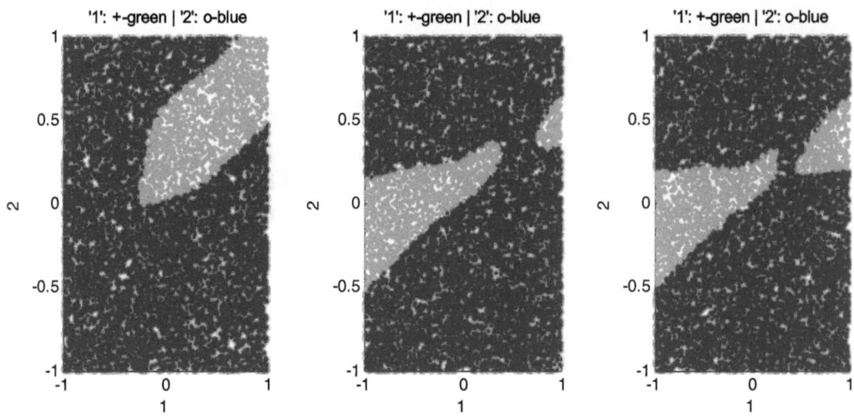

(a) sr = 0 **(b)** sr = 0.3 **(c)** sr = 1

Fig. 3.10 Decision regions in ICA mixtures for different supervision ratios (sr)

effect of increasing outliers is masked, i.e., some observation vectors, which are supposed to be outliers, in fact belong to the generative data model. Including them in training would improve the resulting estimated model.

Finally, Fig. 3.10 shows two class decision regions in one of the generated ICA mixtures that were estimated with different supervision ratios. Figure 3.10a, b represent decision regions for unsupervised training and a low-supervision ratio training, respectively. Figure 3.10c shows the decision regions for supervised training. The accuracy of the decision regions improved with higher supervision since only observation vectors belonging to the generative data model were used in the training stage.

3.5 Conclusions

A novel procedure so-called Mixca for learning the parameters of mixtures of independent component analyzers (mixture matrices, centroids, and source probability densities) has been introduced. The proposed method provides a versatile framework with the following characteristics: no assumptions about the densities of the original sources are required; mixtures with nonlinear dependencies and semi-supervised learning are considered; and any ICA algorithm can be incorporated for updating the model parameters. Considering this last characteristic, a set of variants depending on the embedded algorithm used for ICAMM parameter updating can be defined, e.g., Mixca, Mixca-JADE, Mixca-FastIca, etc. The suitability of application of the proposed technique has been demonstrated in several ICA mixtures and ICA datasets. The non-parametric approach of the procedure clearly yielded better results in source separation than standard ICA algorithms, indicating promising adaptive properties in learning source densities using small sample sizes. In addition, the estimation of multiple ICA parameters of the proposed method provides an important advantage of flexibility over other non-parametric and standard ICA algorithms when data with linear/nonlinear dependencies and complex structures are processed.

The correction of the posterior probability has proven to be useful in the improvement of classification accuracy for ICA mixtures with nonlinear dependencies. This correction is practical for lower and medium dimensions, and for higher dimensions it is constrained by limitations of data availability. The use of few parameters (such as the ICA model parameters) provides efficiency in the representation of the data mixture with the restriction of assuming linear dependencies in the latent variables of the generative data model. However, but this restriction is relaxed to the proposed correction. Thus, the proposed method can be applied to a range of classification problems with data generated from underlying ICA models with residual dependencies, and it may be suitable for the analysis of real-world mixtures.

The role of unlabelled data in training and classification for semi-supervised learning of ICA mixtures has been demonstrated, showing that unlabelled data can degrade the performance of the classifier when they do not fit the generative model of the data. In addition, the fuzziness (the strength of the membership of the data to the classes) of the data mixture contributes to determining the role of the unlabelled data.

References

1. T.W. Lee, M.S. Lewicki, T.J. Sejnowski, ICA mixture models for unsupervised classification of non-gaussian classes and automatic context switching in blind signal separation. IEEE Trans. Pattern Anal. Mach. Intell. **22**(10), 1078–1089 (2000)
2. N.H. Mollah, M. Minami, S. Eguchi, Exploring latent structure of mixture ICA models by the minimum ß-divergence method. Neural Comput. **18**, 166–190 (2005)
3. R. Choudrey, S. Roberts, Variational mixture of bayesian independent component analysers. Neural Comput. **15**(1), 213–252 (2002)
4. A. Leon-García, *Probability and Random Processes for Electrical Engineering* (Addison Wesley, Reading, 1994)
5. A. Salazar, L. Vergara, A. Serrano, J. Igual, A general procedure for learning mixtures of independent component analyzers. Pattern Recogn. **43**(1), 69–85 (2010)
6. B.W. Silverman, *Density Estimation for Statistics and Data Analysis* (Chapman and Hall, London, 1985)
7. S.I. Amari, Natural gradient works efficiently in learning. Neural Comput. **10**, 251–276 (1998)
8. D.T. Pham, Blind separation of instantaneous mixture of sources via an independent component analysis. IEEE Trans. Signal Process. **44**(11), 2768–2779 (1996)
9. A. Samarov, A. Tsybakov, Nonparametric independent component analysis. Bernoulli **10**(4), 565–582 (2004)
10. A. Chen, P.J. Bickel, Efficient independent component analysis. Ann. Stat. **34**(6), 2825–2855 (2006)
11. A. Chen, P.J. Bickel, Consistent independent component analysis and prewhitening. IEEE Trans. Signal Process. **53**(10), 3625–3632 (2005)

12. W. Liu, D.P. Mandic, A. Cichocki, Blind source extraction based on a linear predictor. IET Signal Process. **1**(1), 29–34 (2007)
13. S.I. Amari, T.P. Chen, A. Cichocki, Nonholonomic orthogonal learning algorithms for blind source separation. Neural Comput. **12**, 1463–1484 (2000)
14. T.W. Lee, M. Girolami, T.J. Sejnowski, Independent component analysis using an extended InfoMax algorithm for mixed sub-gaussian and super-gaussian sources. Neural Comput. **11**(2), 417–441 (1999)
15. D. de Ridder, J. Kittler, R.P.W. Duin, Probabilistic PCA and ICA subspace mixture models for image segmentation, in *Proceedings of the British Machine Vision Conference*, Bristol (2000), pp. 112–121
16. A. Hyvärinen, E. Oja, A fast fixed-point algorithm for independent component analysis. Neural Comput. **9**(7), 1483–1492 (1998)
17. J.F. Cardoso, A. Souloumiac, Blind beamforming for non gaussian signals. IEE Proc. F **140**(6), 362–370 (1993)
18. A. Ziehe, K.R. Müller, TDSEP—an efficient algorithm for blind separation using time structure, in *Proceedings of the 8th International Conference on Artificial Neural Networks, ICANN'98, Perspectives in Neural Computing* (1998), pp. 675–680
19. A.J. Bell, T.J. Sejnowski, An information-maximization approach to blind separation and blind deconvolution. Neural Comput. **7**, 1129–1159 (1995)
20. F.R. Bach, M.I. Jordan, Kernel independent component analysis. J. Mach. Learn. Res. **3**, 1–48 (2002)
21. R. Boscolo, H. Pan, Independent component analysis based on nonparametric density estimation. IEEE Trans. Neural Netw. **15**(1), 55–65 (2004)
22. E.G. Learned-Miller, J.W. Fisher, ICA using spacings estimates of entropy. J. Mach. Learn. Res. **4**, 1271–1295 (2003)
23. S. Haykin, *Neural Networks—A comprehensive Foundation*, 2nd edn. (Prentice-Hall, Englewood Cliffs, 1998)
24. A. Hyvärinen, J. Karhunen, E. Oja, *Independent Component Analysis* (Wiley, New York, 2001)
25. H. Lappalainen, A. Honkela, Bayesian nonlinear independent component analysis by multi-layer perceptrons, in *Advances in Independent Component Analysis*, ed. by M. Girolami (Springer, Berlin, 2000), pp. 93–121
26. C.M. Bishop, *Neural Networks for Pattern Recognition* (Oxford University Press, Oxford, 2004)
27. B. Shahshahani, D. Landgrebe, Effect of unlabelled samples in reducing the small sample size problem and mitigating the Hughes phenomenon. IEEE Trans. Geosci. Remote Sens. **32**(5), 1087–1095 (1994)
28. S. Baluja, Probabilistic modelling for face orientation discrimination: learning from labelled and unlabelled data, in *Proceedings of the Neural Information and Processing Systems (NIPS)* (1998), pp. 854–860
29. T. Mitchell, The role of unlabelled data in supervised learning, in *Proceedings of the Sixth Int'l Colloquium Cognitive Science* (1999)
30. K. Nigam, A.K. McCallum, S. Thrun, T. Mitchell, Text classification from labelled and unlabelled documents using EM. Mach. Learn. **39**, 103–134 (2000)
31. I. Cohen, F.G. Cozman, N. Sebe, M.C. Cirelo, T.S. Huang, Semisupervised learning of classifiers: theory, algorithms, and their application to human–computer interaction. IEEE Trans. Pattern Anal. Mach. Learn. **26**(12), 1553–1567 (2004)
32. V. Barnet, T. Lewis, *Outliers in statistical data* (Wiley, New York, 1994)
33. J.C. Bezdek, S.K. Pal, *Fuzzy models for pattern recognition: methods that search for structures in data* (IEEE Press, New York, 1992)
34. R.S. Raghavan, A method for estimating parameters of K-distributed clutter. IEEE Trans. Aerosp. Electron. Syst. **27**(2), 268–275 (1991)
35. D.J. Mackay, *Information theory, inference and learning algorithms* (Cambridge University Press, Cambridge, 2004)

Chapter 4
Hierarchical Clustering from ICA Mixtures

4.1 Introduction

In this chapter, we present a procedure for clustering (unsupervised learning) data from a model based on mixtures of independent component analyzers. Clustering techniques have been extensively studied in many different fields for a long time. They can be organized in different ways according to several theoretical criteria. However, a rough widely accepted classification of these techniques is: hierarchical and partitional clustering; see for instance [1]. Both clustering categories provide a division of the data objects. The hierarchical approach also yields a hierarchical structure from a sequence of partitions performed from singleton clusters to a cluster including all data objects (agglomerative or bottom-up strategy) or vice versa (divisive or top-down strategy). This structure consists of a binary tree (dendrogram) whose leaves are the data objects and whose internal nodes represent nested clusters of various sizes. The whole node of the dendrogram represents the whole data set. The internal nodes describe the extent that the objects are proximal to each other; and the height of the dendrogram usually represents the distance between each pair of objects or clusters, or an object and a cluster.

A review of the clustering algorithms should include the following types of algorithms: hierarchical; squared error-based (vector quantization); mixture density-based; graph theory-based; combinatorial search technique-based; fuzzy; neural network-based; and kernel-based. In addition, some techniques have been developed to tackle sequential, large-scale, and high-dimensional data sets [2]. The advantages of hierarchical clustering include embedded flexibility regarding the level of granularity and the ability to deal with different types of attributes. The disadvantages of hierarchical clustering are the difficulty of scaling up to large data sets, the vagueness of stopping criteria, and the fact that most clustering algorithms cannot recover from poor choices when merging or splitting data points [3].

A. Salazar, *On Statistical Pattern Recognition in Independent Component Analysis Mixture Modelling*, Springer Theses, DOI: 10.1007/978-3-642-30752-2_4,

A proximity or similarity measure is the basis for most clustering algorithms. This measure between clusters at one level in the hierarchy (also referred to as distance) is used to determine which of them will be merged. The distance between two clusters can be estimated between pairs of data objects of each of the clusters or between probabilistic relationships of the data densities of the two clusters. The following general recurrence formula for estimating a function distance $D(*, *)$ was proposed in [4]:

$$D\big(C_l(C_i, C_j)\big) = \alpha_i D(C_l, C_i) + \alpha_j D\big(C_l, C_j\big) + \beta_i D\big(C_i, C_j\big) + \gamma\big|D(C_l, C_i) - D\big(C_l, C_j\big)\big| \tag{4.1}$$

Equation (4.1) describes the distance between a cluster l and a new cluster formed by the merging of two clusters i and j. By manipulating the coefficients $\alpha_i, \alpha_j, \beta$, and γ, several hierarchical algorithms of clustering based on distances between data objects can be derived. Note that if $\alpha_i = \alpha_j = 1/2$, $\beta = 0$, and $\gamma = -1/2$, Eq. (4.1) is $D\big(C_l(C_l, C_j)\big) = \min\big(D(C_l, C_i), D\big(C_l, C_j\big)\big)$ which corresponds to the single linkage method. In the case that $\alpha_i = \alpha_j = \gamma = 1/2$ and $\beta = 0$, Eq. (4.1) becomes $D\big(C_l(C_l, C_j)\big) = \max\big(D(C_l, C_i), D\big(C_l, C_j\big)\big)$, which corresponds to the complete linkage method [5].

The probabilistic approaches to hierarchical clustering consider model-based criteria or Bayesian hypotheses to decide on merging clustering rather than using an ad-hoc distance metric. Basically, there are two approaches to derive the hierarchy: hierarchical generative modelling of the data or hierarchical ways of organizing nested clusters. Methods of the first approach include the following hierarchical generative models, for instance: Gaussian- [6], diffusion- [7], and mutation process-based [8]. The first two methods can be used for inference, and the last one can be used for semi-supervised learning. In [9], an agglomerative algorithm to merge of Gaussian mixtures is presented. It considers a virtual sample generated from the model at a level and uses EM to find the expressions for the mixture model parameters for the next level that best explain the virtual sample. Methods of the second approach include: agglomerative model merging, which is based on marginal likelihoods in the context of HMM [10]; a method to compute the marginal likelihood for c and $c - 1$ clusters for use in an agglomerative algorithm [11]; clustering of multinomial feature vector data considering subsets of features having common distributions [12]; probabilistic abstraction hierarchies in which each node contains a probabilistic model with the most similar models being neighbouring nodes (estimated by a distance function) [13]; agglomerative algorithm for merging time series based on greedily maximizing marginal likelihood [14, 15]; and using marginal likelihoods to decide which clusters to merge, when to stop, and when to avoid overfitting by testing a Bayesian hypothesis [16]. The model of this last method can be used to compute the predictive distribution of a test point and the probability of it belonging to any of the existing clusters in the tree.

Work on clustering that is related with hierarchies that are derived from ICA can be found in a hierarchical latent variable model for data visualization proposed in [17]. In this model, the form of the latent variable model is closely related to probabilistic principal component analysis (PPCA) [18, 19]. The construction of the hierarchical tree proceeds top-down. At the top level of the hierarchy, a single visualization plot corresponding to a single model is defined. This model is partitioned into "clusters" at the second level of the hierarchy considering a probabilistic mixture of latent variable models. Subsequent levels, which are obtained using nested mixture representations, provide successively refined models of the data set [17]. ICA model-based hierarchies have also been explored. For instance, in [20], a method for capturing nonlinear dependencies in natural images for image segmentation and denoising is presented. It makes use of lower level linear ICA representation and a subsequent mixture of Laplacian distributions for learning the nonlinear dependencies.

The method proposed in this chapter corresponds to hierarchical clustering of agglomerative type. It starts from a set of ICA mixture parameters that are extracted from the data using a learning process as explained in Chap. 3. Each cluster at the first level of the hierarchy is characterized with the parameters of a single ICA model. These parameters (mixture matrices, bias vectors, and source probability density functions) are used to estimate the proximities between clusters pairwise using the Kullback–Leibler distance [21]. The pdf of the sources is estimated using a non-parametric kernel-based density. During the merging of the clusters, the entropy and cross-entropy of the sources have to be estimated. This cannot be obtained analytically, and thus an iterative suboptimal approach is applied using a numerical approximation from the training data.

The structure of several ICA subspaces at the bottom level of the hierarchy allows non-gaussian mixtures to be modelled. The independence relations between the hidden variables at this lowest level are relaxed at higher levels of the hierarchy allowing more flexible modelling. The subspaces constructed by the method at intermediate levels of the hierarchy represent different degrees of dependence of the variables. Thus, this work can be related to tree-dependent component analysis (TCA), which finds "clusters" of components such that the components are dependent within a cluster and independent between clusters [22]. Topographic independent component analysis (TICA) is another method that considers the residual dependence after ICA. This method defines a distance between two components using higher-order correlations, and it is used to create a topographic representation [23].

The clustering procedure was tested with simulations and two applications. The simulations considered ICA mixtures of diverse kinds of densities: uniform, K-type [24], Laplacian, and Rayleigh. The quality of the clustering (selection of number of clusters in the dendrogram) was tested with the partition and partition entropy coefficients [25]. There are many practical applications of statistical learning where it is useful to characterize data hierarchically. We selected real image processing since it is classical and the results can be easily interpreted. The objectives were real object recognition and image segmentation.

4.2 Problem Statement and Distance Between ICA Clusters

The conditional probability density of an observation vector $\mathbf{X}$ for cluster $C_k^h, k = 1, 2, \ldots, K - h + 1$ in layer $h = 1, 2, \ldots, K$ is $p\left(\mathbf{x}/C_k^h\right)$. At the first level, $h = 1$, it is modelled by the K-ICA mixtures, i.e., $p\left(\mathbf{x}/C_k^1\right)$. is:

$$p\left(\mathbf{x}/C_k^1\right) = \left|\det \mathbf{A}_k^{-1}\right| p(\mathbf{s}_k), \ \mathbf{s}_k = \mathbf{A}_k^{-1}(\mathbf{x} - \mathbf{b}_k) \tag{4.2}$$

At each consecutive level, two clusters are merged according to some minimum distance measure until only one cluster is reached at level $h = K$.

For the distance measure, we use the symmetric Kullback–Leibler divergence between the ICA mixtures, which is defined for the clusters u, v by:

$$D_{KL}\left(C_u^h, C_v^h\right) = \int p\left(\mathbf{x}/C_u^h\right) \log \frac{p\left(\mathbf{x}/C_u^h\right)}{p\left(\mathbf{x}/C_v^h\right)} d\mathbf{x} + \int p\left(\mathbf{x}/C_v^h\right) \log \frac{p\left(\mathbf{x}/C_v^h\right)}{p\left(\mathbf{x}/C_u^h\right)} d\mathbf{x} \tag{4.3}$$

For layer $j = 1$, from Eq. (4.3), we can obtain the following:

$$\begin{aligned} D_{KL}(C_u, C_v) &= D_{KL}(p_{\mathbf{x}_u}(\mathbf{x}) // p_{\mathbf{x}_v}(\mathbf{x})) \\ &= \int p_{\mathbf{x}_u}(\mathbf{x}) \log \frac{p_{\mathbf{x}_u}(\mathbf{x})}{p_{\mathbf{x}_v}(\mathbf{x})} d\mathbf{x} + \int p_{\mathbf{x}_v}(\mathbf{x}) \log \frac{p_{\mathbf{x}_v}(\mathbf{x})}{p_{\mathbf{x}_u}(\mathbf{x})} d\mathbf{x} \end{aligned} \tag{4.4}$$

For brevity, we write $p_{\mathbf{x}}(\mathbf{X}) = p\left(\mathbf{x}/C_u^1\right)$ and omit the superscript $h = 1$. For simplicity, we impose the independence hypothesis and we suppose that both clusters have the same number of sources M:

$$\begin{aligned} p_{\mathbf{x}_u}(\mathbf{X}) &= \frac{\prod_{i=1}^{M} p_{s_{u_i}}(s_{u_i})}{|\det \mathbf{A}_u|}, \quad s_{u_i} = \mathbf{A}_{u_i}^{-1}(\mathbf{X} - \mathbf{b}_{u_i}) \\ p_{\mathbf{x}_v}(\mathbf{X}) &= \frac{\prod_{j=1}^{M} p_{s_{v_j}}(s_{v_j})}{|\det \mathbf{A}_v|}, \quad s_{v_i} = \mathbf{A}_{v_i}^{-1}(\mathbf{X} - \mathbf{b}_{v_i}) \end{aligned} \tag{4.5}$$

4.3 Merging ICA Clusters with Kernel-Based Source Densities

The pdf of the sources is approximated by a non-parametric kernel-based density for both clusters:

$$p_{s_{u_i}}(s_{u_i}) = \sum_{n=1}^{N} a e^{-\frac{1}{2}\left(\frac{s_{u_i} - s_{u_i}(n)}{h}\right)^2}, \quad p_{s_{v_j}}(s_{v_j}) = \sum_{n=1}^{N} a e^{-\frac{1}{2}\left(\frac{s_{v_j} - s_{v_j}(n)}{h}\right)^2}. \tag{4.6}$$

where, again for simplicity, we have assumed the same kernel function with the parameters a, h for all the sources and the same number of samples N for each one. Note that this corresponds to a mixture of Gaussian models where the number of Gaussians is maximum (one for every observation) and the parameters are equal. Reducing the pdf of the sources to a standard mixture of Gaussians with a different number of components and priors for each source does not help in computing the Kullback–Leibler distance because there is no analytical solution for it. Therefore, we prefer to maintain the non-parametric approximation of the pdf in order to model more complex distributions than a mixture of a small finite number of Gaussians, such as three or four.

The symmetric Kullback–Leibler distance between the clusters u, v can be expressed as:

$$D_{KL}(p_{\mathbf{x}_u}(\mathbf{x})//p_{\mathbf{x}_v}(\mathbf{x})) = -H(\mathbf{x}_u) - H(\mathbf{x}_v) - \int p_{\mathbf{x}_u}(\mathbf{x}) \log p_{\mathbf{x}_v}(\mathbf{x})\, d\mathbf{x} - \int p_{\mathbf{x}_v}(\mathbf{x}) \log p_{\mathbf{x}_u}(\mathbf{x})\, d\mathbf{x} \quad (4.7)$$

where $H(\mathbf{x})$is the entropy, which is defined as $H(\mathbf{x}) = -E[\log p_{\mathbf{x}}(\mathbf{x})]$. To obtain the distance, we have to calculate the entropy for both clusters as well as the cross-entropy terms $E_{\mathbf{x}_v}[\log p_{\mathbf{x}_u}(\mathbf{x})]$, $E_{\mathbf{x}_u}[\log p_{\mathbf{x}_v}(\mathbf{x})]$

The entropy for the cluster u can be calculated through the entropy of the sources of that cluster taking into account the linear transformation of the random variables and their independence Eq. (4.5):

$$H(\mathbf{x}_u) = \sum_{i=1}^{M} H(s_{u_i}) + \log|\det \mathbf{A}_u| \quad (4.8)$$

The entropy of the sources cannot be analytically calculated. Instead, we can obtain a sample estimate of $\hat{H}(s_{u_i})$ using the training data. Denote the i-th source obtained for the cluster u by $\{s_{u_i}(1), s_{u_i}(2), \ldots, s_{u_i}(Q_i)\}$. The entropy can be approximated as follows:

$$\begin{aligned}\hat{H}(s_{u_i}) &= -\hat{E}\left[\log p_{s_{u_i}}(s_{u_i})\right] = -\frac{1}{Q_i}\sum_{n=1}^{Q_i} \log p_{s_{u_i}}(s_{u_i}(n)) \\ p_{s_{u_i}}(s_{u_i}(n)) &= \sum_{l=1}^{N} a e^{-\frac{1}{2}\left(\frac{s_{u_i}(n)-s_{u_i}(l)}{h}\right)^2}\end{aligned} \quad (4.9)$$

The entropy of $H(\mathbf{x}_v)$ is obtained analogously:

$$\begin{aligned}H(\mathbf{x}_v) &= \sum_{i=1}^{M} H(s_{v_i}) + \log|\det \mathbf{A}_v| = \sum_{i=1}^{M} \hat{H}(S_{v_i}) + \log|\det \mathbf{A}_v| \\ \hat{H}(S_{v_i}) &= -\frac{1}{Q_i}\sum_{n=1}^{Q_i} \log p_{s_{vi}}(s_{v_i}(n)),\ p_{s_{vi}}(s_{v_i}(n)) = \sum_{l=1}^{N} a e^{-\frac{1}{2}\left(\frac{s_{v_i}(n)-s_{v_i}(l)}{h}\right)^2}\end{aligned} \quad (4.10)$$

with $\hat{H}(S_{v_i})$defined analogously to Eq. (4.9).

Other possible approximations are available, for example to use synthetic data produced from the distributions of the sources instead of the data used to learn the parameters of the ICA mixture model.

Once the entropy is computed, we have to obtain the cross-entropy terms.

$$\begin{aligned} E_{\mathbf{x}_u}[\log p_{\mathbf{x}_u}(\mathbf{x})] &= \int p_{\mathbf{x}_u}(\mathbf{x}) \log p_{\mathbf{x}_v}(\mathbf{x}) d\mathbf{x} = \int p_{\mathbf{x}_u}(\mathbf{x}) \log \frac{\prod_{i=1}^{M} p_{s_{v_i}}(s_{v_i})}{|\det \mathbf{A}_v|} d\mathbf{x} \\ E_{\mathbf{x}_v}[\log p_{\mathbf{x}_u}(\mathbf{x})] &= \int p_{\mathbf{x}_v}(\mathbf{x}) \log p_{\mathbf{x}_u}(\mathbf{x}) dx = \int p_{\mathbf{x}_v}(\mathbf{x}) \log \frac{\prod_{i=1}^{M} p_{s_{u_i}}(s_{u_i})}{|\det \mathbf{A}_u|} d\mathbf{x} \end{aligned} \tag{4.11}$$

Considering the relationships $\mathbf{x} = \mathbf{A}_u\mathbf{s}_u + \mathbf{b}_u$, $\mathbf{x} = \mathbf{A}_v\mathbf{s}_\mathbf{v} + \mathbf{b}_v$ and thus $\mathbf{s}_v = \mathbf{A}_v^{-1}$ $(\mathbf{A}_u\mathbf{s}_u + \mathbf{b}_u - \mathbf{b}_v)$, we obtain for the first cross-entropy in Eq. (4.11):

$$\int p_{\mathbf{x}_u}(\mathbf{x}) \log p_{\mathbf{x}_v}(\mathbf{x})\, d\mathbf{x} = \int p_{\mathbf{s}_u}(\mathbf{s}) \log \frac{\prod_{i=1}^{M} \sum_{n=1}^{N} a e^{-\frac{1}{2}\left(\frac{s_{v_i} - s_{v_i}(n)}{h}\right)^2}}{|\det \mathbf{A}_v|} d\mathbf{s} \tag{4.12}$$

with s_{v_i} being the i-th element of the vector $\mathbf{s}_v$, i.e.,$s_{v_i} = \left[\mathbf{A}_v^{-1}(\mathbf{A}_u\mathbf{s} + \mathbf{b}_u - \mathbf{b}_v)\right]_i$

Using Eq. (4.12), by applying the independence of the sources for the cluster u, we can obtain:

$$\begin{aligned} \int p_{\mathbf{x}_u}(\mathbf{x}) \log p_{\mathbf{x}_v}(\mathbf{x})\, d\mathbf{x} &= -\log|\det \mathbf{A}_V| + \int \prod_{i=1}^{M} \sum_{n=1}^{N} a e^{-\frac{1}{2}\left(\frac{s_{v_i} - s_{v_i}(n)}{h}\right)^2} \\ d\mathbf{s} &= -\log|\det \mathrm{A}_V| + \sum_{i=1}^{M} \int p_{s_{u_M}}(s_M) ds_M \ldots \int p_{s_{u_1}}(s_1) \log \sum_{n=1}^{N} a e^{-\frac{1}{2}\left(\frac{s_{v_i} - s_{v_i}(n)}{h}\right)^2} ds_1 \end{aligned} \tag{4.13}$$

Again, there is no analytical solution to Eq. (4.13), so we have to use numerical alternatives to approximate the cross-entropy. Following the same idea as above with the entropy, we can use the data corresponding to every source for cluster u in order to approximate the expectation of Eq. (4.13). Assuming that we have or can generate Q_i observations according to distribution $p_{s_{u_i}}(s_i)$ $i = 1, \ldots, M$, we can estimate

$$\begin{aligned} &\int p_{s_{u_M}}(s_M) ds_M \ldots \int p_{s_{u_1}}(s_1) \log \sum_{n=1}^{N} a e^{-\frac{1}{2}\left(\frac{s_{v_i} - s_{v_i}(n)}{h}\right)} \\ &ds_1 \approx \frac{1}{\prod_{i=1}^{M} Q_i} \sum_{s_M=1}^{Q_M} \ldots \sum_{s_1=1}^{Q_1} \log \sum_{n=1}^{N} a e^{-\frac{1}{2}\left(\frac{s_{v_i} - s_{v_i}(n)}{h}\right)^2} \end{aligned} \tag{4.14}$$

with $s_{v_i} = (\mathbf{Bs})_i + \mathbf{c}_i, \mathbf{B} = \mathbf{A}_v^{-1}\mathbf{A}_u, \mathbf{c}_i = \mathbf{A}_v^{-1}(\mathbf{b}_u - \mathbf{b}_v), \mathbf{s} = [s_1(k), \ldots, s_M(l)]^T, k \in [1, Q_1], \ldots l \in [1, Q_M]$. Of course, the other term in Eq. (4.11) is obtained in a similar way, considering that now the expectations are obtained by averaging the pdf of the sources of the other cluster.

Taking into account all the terms in Eq. (4.4), the symmetrical Kullback–Leibler distance between clusters u, v can be computed numerically from the samples following the corresponding distribution $\{s_{u_i}(1), s_{u_i}(2), \ldots, s_{u_i}(Q)\}, \quad i = 1, \ldots, M$, $\{s_{v_j}(1), s_{v_j}(2), \ldots, s_{v_j}(Q)\} \quad j = 1, \ldots, M$ (we assume that the number of samples per source is the same for all of them). The computation is summarized as follows:

$$\begin{aligned}
D_{KL}(p_{x_u}(x)//p_{x_v}(x)) &= -\sum_{i=1}^{M}\hat{H}(S_{u_i}) - \sum_{j=1}^{M}\hat{H}(S_{v_j}) - \sum_{i=1}^{M}\hat{H}(\mathbf{S}_v, S_{u_i}) - \sum_{j=1}^{M}\hat{H}(\mathbf{S}_u, S_{v_j}) \\
\hat{H}(S_{u_i}) &= \frac{1}{Q}\sum_{n=1}^{Q}\log p_{s_{u_i}}(S_{u_i}(n)),\ p_{s_{u_i}}(S_{u_i}(n)) = \sum_{l=1}^{N} a e^{-\frac{1}{2}\left(\frac{s_{u_i}(n)-s_{u_i}(l)}{h}\right)^2} \\
\hat{H}(S_{v_j}) &= \frac{1}{Q}\sum_{n=1}^{Q}\log p_{s_{v_j}}(S_{v_j}(n)),\ p_{s_{v_j}}(S_{v_j}(n)) = \sum_{l=1}^{N} a e^{-\frac{1}{2}\left(\frac{s_{v_j}(n)-s_{v_j}(l)}{h}\right)^2} \\
\hat{H}(\mathbf{S}_v, S_{u_i}) &= \frac{1}{Q^M}\sum_{s_{v1}=1}^{Q}\cdots\sum_{s_{vM}=1}^{Q}\log\sum_{n=1} Nae^{-\frac{1}{2}\left(\frac{[A_u^{-1}(A_v s_v + b_v - b_u)]_j - s_{u_i}(n)}{h}\right)^2} \\
\hat{H}(\mathbf{S}_u, S_{v_j}) &= \frac{1}{Q^M}\sum_{s_{u1}=1}^{Q}\cdots\sum_{s_{uM}=1}^{Q}\log\sum_{n=1} Nae^{-\frac{1}{2}\left(\frac{[A_v^{-1}(A_u s_u + b_u - b_v)]_j - s_{v_j}(n)}{h}\right)^2}
\end{aligned} \tag{4.15}$$

As can be observed, the similarity between clusters depends not only on the similarity between the bias terms, but also on the similarity between the distributions and the mixing matrices. The computations can also be easily extended to the case where the number of sources in every class is not the same. In the case that the distributions are approximated by just a single Gaussian (keeping in mind that the ICA problem reduces to the PCA problem since there is an indetermination defined by an orthogonal matrix that is not identifiable) and the distance is obtained analytically for the first level of the hierarchy, the distance between two multivariate normal distributions of dimension M $p_u(\mathrm{x}) = N(\mu_u, \Sigma_u)$, $p_v(\mathrm{x}) = N(\mu_v, \Sigma_v)$ would be

$$\begin{aligned}
D_{KL}(p_u(\mathrm{x})//p_v(\mathrm{x})) = tr\left(\Sigma_u\Sigma_v^{-1}\right) + tr\left(\Sigma_v\Sigma_u^{-1}\right) - 2M \\
+ tr\left[\left(\Sigma_u^{-1}\Sigma_v^{-1}\right)(\mu_u - \mu_v)(\mu_u - \mu_v)^T\right]
\end{aligned} \tag{4.16}$$

where $tr[\mathrm{A}]$ is the trace of matrix $\mathbf{A}$.

Once the distances are obtained for all the clusters, the two clusters with minimum distance are merged at a certain level. This is repeated in each step of the hierarchy until one cluster at the level $h = k$ is reached. To merge a cluster at level h, we can calculate the distances from the distances of level $h - 1$. Suppose that from level $h - 1$ to h the clusters C_u^{h-1}, C_v^{h-1} are merged in cluster C_w^h. Then, the density for the merged cluster at level h is:

$$p_h\left(\mathbf{x}/C_w^h\right) = \frac{p_{h-1}\left(C_u^{h-1}\right)p_{h-1}\left(\mathbf{x}/C_u^{h-1}\right) + p_{h-1}\left(C_v^{h-1}\right)p_{h-1}\left(\mathbf{x}/C_v^{h-1}\right)}{p_{h-1}\left(C_u^{h-1}\right) + \left(C_v^{h-1}\right)} \tag{4.17}$$

where $p_{h-1}\left(C_u^{h-1}\right)$, $p_{h-1}\left(C_v^{h-1}\right)$ are the priors or proportions of the clusters u, v at level h − 1. The rest of the terms are the same in the mixture model at level h as at level $h - 1$. The only difference from one level to the next one in the hierarchy is that there is one cluster less and the prior for the new cluster is the sum of the priors of its components and the density the weighted average of the densities that are merged to form it. Therefore, the estimation of the distance at level h can be done easily starting from the distances at level $h - 1$ and so on until level $h = 1$. Consequently, we can calculate the distances at level h from a cluster C_z^h to a merged cluster C_w^h that was obtained by the agglomeration of clusters C_u^{h-1}, C_v^1 at level $h - 1$ as the distance to its components weighted by the mixing proportions:

$$D_h\left(p_h\left(\mathrm{x}/C_w^h\right)//p_h\left(\mathrm{x}/C_z^h\right)\right) = \frac{p_{h-1}\left(C_u^{h-1}\right)\cdot D_{h-1}\left(p_{h-1}\left(\mathrm{x}/C_u^{h-1}\right)//p_{h-1}\left(\mathrm{x}/C_z^{h-1}\right)\right) + p_{h-1}\left(C_v^{h-1}\right)\cdot D_{h-1}\left(p_{h-1}\left(\mathrm{x}/C_v^{h-1}\right)//p_{h-1}\left(\mathrm{x}/C_z^{h-1}\right)\right)}{p_{h-1}\left(C_u^{h-1}\right) + p_{h-1}\left(C_v^{h-1}\right)} \tag{4.18}$$

As at level 1, we can obtain the decision rule to assign a new data to a cluster in the hierarchy at level h by applying Bayes' theorem:

$$\arg\max_{C_k} p_h(C_k/\mathrm{x}) = \frac{p_h(\mathrm{x}/C_k)\,p_h(C_k)}{\sum_{i=1}^{K-h+1} p_h(\mathrm{x}/C_i)\,p_h(C_i)} \tag{4.19}$$

4.3.1 ICAMM-Based Hierarchical Clustering Algorithm

The summary of the ICAMM-based hierarchical clustering algorithm is presented in Table 4.1.

4.4 Simulations

The algorithm above was tested with several simulated data of ICA mixtures with the following source distributions: Laplacian, uniform, K-type, and Rayleigh. The mixtures were generated for two independent variables and bias terms for a number of 200 observation vectors. Figure 4.1 shows some of the source distributions used in the ICA mixtures.

Table 4.1 ICAMM-based hierarchical clustering algorithm

0. Estimate the ICA parameters for each of the clusters at the bottom of the hierarchy at level $h = 1$
   $\mathrm{A}_k, \mathrm{s}_k, \mathrm{b}_k, \quad k = 1, \ldots K,$
   use the Mixca algorithm in Chap. 3.
1. Estimate pairwise distance for every pair of clusters
   $\mathrm{x}^1, \mathrm{x}^2, \ldots, \mathrm{x}^N$, use Eq. (4.15) to estimate $\hat{H}(s_{u_i}), \hat{H}(s_{v_j}), \hat{H}(s_v, s_{u_i}), \hat{H}(s_u, s_{v_j})$ and non-parametric kernel-based estimation of the source pdf.
2. Select the pair of clusters with minimum distance to merge.
3. Estimate the density for the merged cluster $h = h + 1$
   $$p_h(\mathrm{x}/C_w^h) = \frac{p_{h-1}(C_u^{h-1})p_{h-1}(\mathrm{x}/C_u^{h-1}) + p_{h-1}(C_v^{h-1})p_{h-1}(\mathrm{x}/C_v^{h-1})}{p_{h-1}(C_u^{h-1}) + p_{h-1}(C_v^{h-1})},$$
   where $p_{h-1}(C_u^{h-1}), p_{h-1}(C_v^{h-1})$ are the priors of clusters u,v at level $h-1$.
4. Calculate the distance from a cluster C_z^h to a merged cluster C_w^h
   $C(\mathbf{W}) = \hat{I}_{\lambda F}(K_1, \ldots, K_m),$
   use $\arg\max_{C_k} p_h(C_k/\mathrm{x}) = \dfrac{p_h(\mathrm{x}/C_k)p_h(C_k)}{\sum_{i=1}^{K-h+1} p_h(\mathrm{x}/C_i)p_h(C_i)}$ to assign a new data to a cluster in the hierarchy
5. Repeat steps 2–4 until one cluster is obtained at the top of the hierarchy

We present simulation results that illustrate the properties of the hierarchical algorithm proposed. We start by a simple example that illustrates how the algorithm can be used to estimate hierarchical mixtures. Figure 4.2 shows ten different ICA mixtures generated from sources with Laplacian (L) and uniform (U) distributions: 1. ◇ (U), 2. ∇ (U), 3. Δ (L), 4. o (L), 5. > (U), 6. + (U), 7. < (L), 8. □ (L), 9. * (U), 10. × (L). The hierarchical description of the data is shown at three resolutions ($h = 1, 5, 9$). Notice how the mixture hierarchy naturally captures the various levels of structure exhibited by the data. Figure 4.2a shows an example of how the algorithm first learned the basis vectors and bias terms for each class at level h = 1 and then used these estimations to create a hierarchical tree of the clusters. Figure 4.2b shows the merged clusters at level h = 5, where six clusters remain. The closest clusters are merged. In Fig. 4.2c, there are only two clusters remaining. Note how not only the relative bias between clusters is used to merge them, but also similar densities are merged together. Figure 4.3 shows the resulting dendrogram with the distances at which the clusters have been merged. The higher the level of the hierarchy, the larger the distance required to merging the clusters, and the independence assumption of the bottom level becomes weaker.

In order to obtain a good performance of the agglomerative clustering, the bandwidth parameter (h in Eq. (4.17)) has to be adjusted in order to keep the distances from being thresholded to one value. There is a compromise between the resolution of the distances and their range. Thus, we chose a high bandwidth.

In order to determine a quality criterion that allows the optimum hierarchical level to be determined, the partition $\left(PC = -\frac{1}{N}\sum_{i=1}^{N}\sum_{j=1}^{nc} u_{ij}^2\right)$ and partition entropy

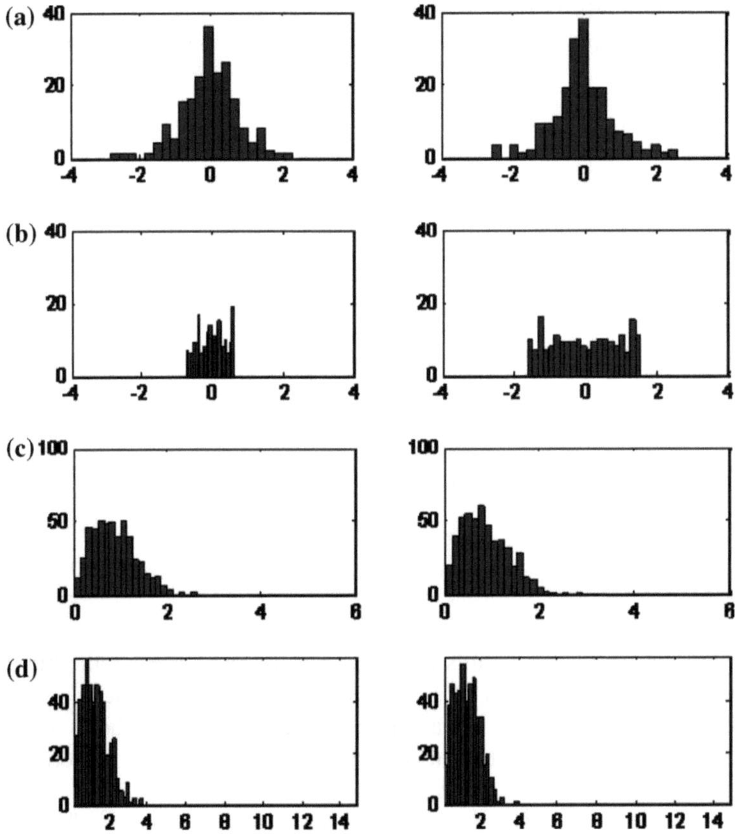

Fig. 4.1 Some distributions of the sources used in hierarchical clustering. **a** Laplacian, **b** uniform, **c** K-type m = 10, **d** Rayleigh

coefficient $\left(PE = -\frac{1}{N}\sum_{i=1}^{N}\sum_{j=1}^{nc} u_{ij} \cdot \log_a\left(u_{ij}\right)\right)$ were estimated [25]. The partition coefficient and the partition entropy both tend towards monotone behaviour depending on the number of clusters. Therefore, to find the optimum number of clusters, the number where the entropy value lies below the rising trend and where the value for the partition coefficient lies above the falling trend is selected. The point of the curve of all the connected values can be identified as a kink ("elbow criterion") where the optimum number of clusters is located. Figure 4.4 shows the evolution of the above coefficients through the clustering levels of data of Figure 4.2.

The optimum partitioning of the clusters applies at that point of the dendrogram that has a value of h to obtain the highest cluster differentiation (maximum of inter-cluster mean distances) with good homogeneity within cluster members (minimum of distances between members of the clusters and centroids). From

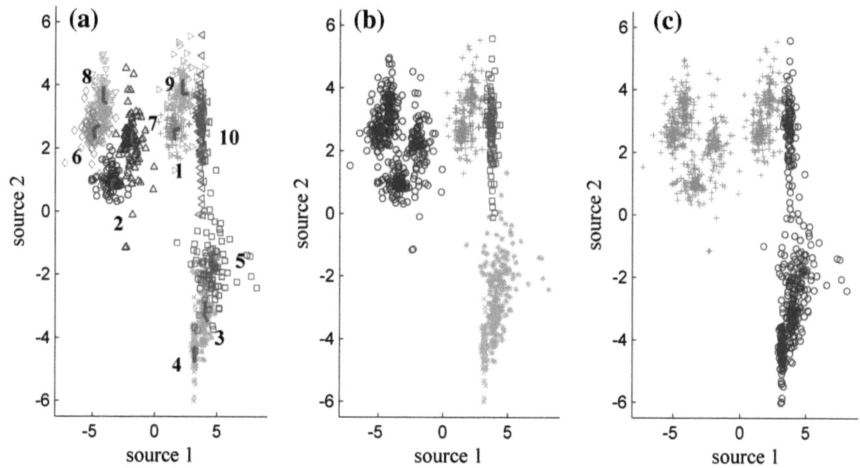

Fig. 4.2 Hierarchical classification from ten ICA mixtures (Laplacian and uniform source distributions). The bottom-up construction of higher-level clusters is indicated. **a** Data and basis vectors for each ICA mixture at level $h = 1$, **b** clusters at level $h = 5$, **c** clusters at level $h = 9$

Fig. 4.3 Dendrogram with KL-distances between clusters at each merging level

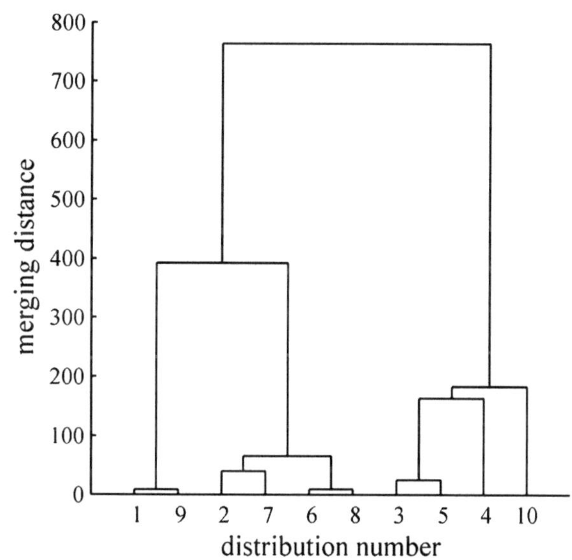

Fig. 4.4 and following the criterion of coefficients PC and PE explained above, the optimum number of clusters was obtained at level $h = 5$ (i.e., grouping the data in 6 clusters as shown in Fig. 4.2b).

The first step of the proposed hierarchical algorithm is to estimate the parameters of the ICA mixtures that will determine the clusters at the lowest level of the hierarchy. Thus, errors in the source densities of the model at this level will propagate to higher levels producing erroneous groupings. In order to measure the

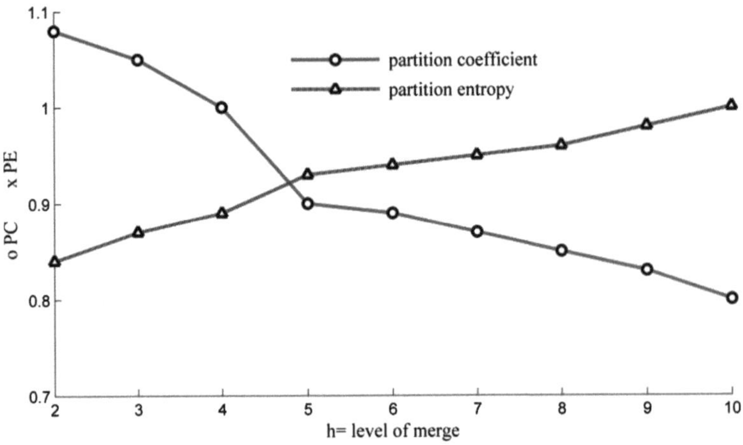

Fig. 4.4 Measure of clustering quality as a function of the merging level

performance of ICAMM parameter estimation, which is the basis of the hierarchical clustering, 400 Montecarlo simulations were made with toy data such as the distributions shown in the examples of Fig. 4.1. The algorithm was executed varying the number of mixtures, kinds of source distributions, and supervision ratio in the learning stage of the parameters.

Figure 4.5 shows the classification error for different numbers of ICA mixtures (hierarchical levels). Note that the percentage of error rapidly decreases as long as there is a little supervision and the densities are estimated properly. For the unsupervised case, the error percentage is higher when the number of ICA mixtures is greater, with the highest being (0.15) for $K = 6$ and the lowest being (0.096) for $K = 3$. However, this difference in classification accuracy is less significant for all the semi-supervised cases. This shows the advantage of employing a semi-supervised and non-parametric density estimation algorithm in the first step of the hierarchical algorithm.

Figure 4.6 shows an example of application in simulations of the proposed hierarchical algorithm. The ICA mixtures are Laplacian source distributions and combinations of different source distributions (Laplacian, uniform, K-type $m = 10$, Rayleigh) for the first and second example, respectively. The dendrograms estimated by the proposed method and the single linkage method are included for comparison. There are significant differences between the two methods of agglomerative clustering. The proposed method is based on probabilistic distances between groups of data, while the single linkage method uses the distance between pairs of data objects (for these examples, we used Euclidean distance). In addition, the first method employs the parameters of the underlying model estimated at the lowest hierarchical level, while the second one does not. This could be important depending on the data structure, i.e., if the data follows and ICA mixture model. These differences determine variations in the dendrograms delivered for these two methods. Thus, it is expected that the merging

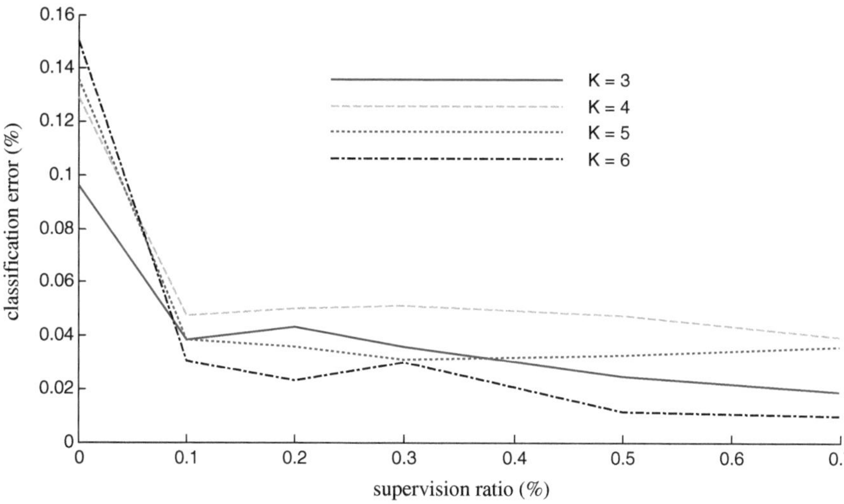

Fig. 4.5 Classification error as a function of the mixture clusters and the supervision ratio

sequence and distance be different, and thus, the constructed intermediate subspaces (with their localized patterns) are different. As a consequence of this, the optimum clustering will be different for these two methods. Theoretically, model-based clustering allows general patterns to be learned, while clustering using only data objects can be biased to learn particular patterns of the given data examples.

The dendrograms of Fig. 4.6 are equivalent at the penultimate level of the hierarchy (they found the two principal groupings of the data); however, there are several differences in the intermediate hierarchy levels. For instance, the first three mergings for the proposed method were between clusters: 5–6, 1–3, and 7–8; while for the single linkage method the mergings were: 1–2, (1–2)–3, 7–10. It is clear that the shape of the data densities favours the selection of the clusters to be merged in the case of the proposed method, instead of the mass of data that is more important for the single linkage method. Note that the sequence of merging for the first method proceeds including clusters of the two larger zones of the data, while the single linkage method focuses on only one zone in the first two mergings. Thus, the shape of the subspaces will be quite different for the two methods at the fourth level of the hierarchy.

4.5 Real Data Analysis: Image Processing

Local edge detectors can be extracted from natural scenes by ICA algorithms [26, 27, 28]. ICA can be used for estimating features from natural images for sparse coding interpretation, i.e., the data vector is represented using a set of basis vectors so that only a small number of basis vectors are activated at the same time. These vectors

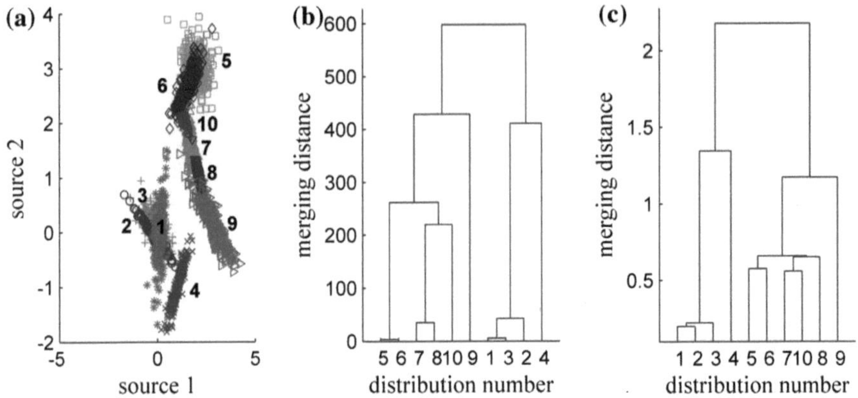

Fig. 4.6 Hierarchical classification from ten ICA mixtures (Laplacian source distributions). **a** Data, **b** dendrogram estimated by the proposed method, **c** dendrogram estimated by the single-linkage algorithm

have been related with the detection of borders in natural images [26]. Therefore, basis functions have a physical relation with objects and can be used to measure the similarity between objects based on ICA decomposition. In image patch decomposition, the set of independent components is larger than what can be estimated at one time, and the result at one time is an arbitrarily chosen subset [23]. Nevertheless, ICA has been applied successfully in several image applications; see for instance [29]. Neurophysiological primary visual cortex activities have also been related with the detection of borders in natural images. This has allowed theoretical dynamic models of the abstraction process in living beings to be proposed (which proceeds from the visual cortex to higher-level abstraction) [30].

This section is intended to process the parameters of ICA mixtures in order to obtain hierarchical structures from the basis function level (edges) to higher levels of clustering. Specifically, the proposed clustering algorithm is applied to image analysis obtaining promising results in discerning object similarity (object recognition task) and suitable levels of hierarchies by processing patches of natural images. Thus, the feedforward process that is featured by the hierarchical clustering method can be related with the process of abstraction.

4.5.1 Real Object Recognition

For the object recognition task, the Columbia Object Image Library -100 (COIL-100) database was used [31]. The database consists of a large number of images of objects over a black background (views taken from different angles). The size of the images is 128 × 128 pixels. Some of the object images are very similar to others depending on the angle from which the picture of the object was made.

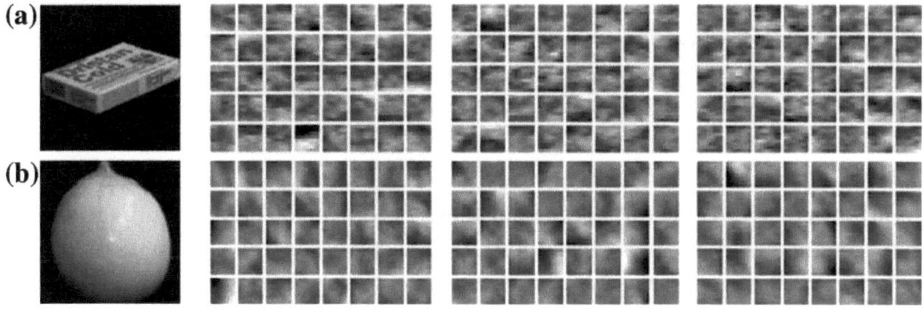

Fig. 4.7 Two groups of basis functions corresponding to two different objects. The basis functions at top are from a small box and the basis functions at bottom are from an onion

In all the experiments each image was first converted to greyscale and linearly normalized so that the pixels had zero mean and unit variance. The first test was to compare the basis functions of different objects in COIL-100 database. A total of 20 images for each one of eight selected objects were randomly taken from the database. A total of 2,000 image patches (windows) of 8×8 pixels were randomly taken for each object. From each patch the local mean was subtracted. These data were used to estimate the basis functions previous to a whitening process using PCA, with a reduction from 64 features to 40 components. This procedure is explained in detail in [26, 32]. The basis functions were then calculated with the Mixca algorithm that was performed one time per each object data set in order to estimate the parameters of three classes. Supervised training and the Laplacian prior was used in order to estimate the source pdf's. The estimated basis functions were converted to the original feature space using the dewhitening matrix previously estimated by PCA. Figure 4.7 shows the 40 basis functions of 8×8 corresponding to the three class parameters estimated for two of the objects: a box with an inscribed label (Fig. 4.7a) and an onion (Fig. 4.7b). The similarities and differences between the functions of each object can be observed in this figure: for instance, the lower frequency in the pattern corresponding to a natural object (the onion) versus the high frequency in the pattern of an artificial object (the box).

The obtained basis functions were used to measure the distance between classes by estimating the symmetric Kullback–Leibler (KL) distance from the mixture matrices previously calculated. The KL distances revealed that basis functions can find the similarity (short distances) between classes corresponding to the same object (intra-object), and the difference (long distances) between classes of different objects (inter-object). The KL distances corresponding to the two objects in Fig. 4.7 are shown in Table 4.2.

Experiments to create a hierarchical classification of objects were also performed using the data set of the experiment above. The Mixca algorithm was used to estimate the parameters of eight classes (one class per object). These ICA parameters build the lowest level of the hierarchy. The intermediate levels of the hierarchy were then created by applying the agglomerative clustering algorithm explained in Sect. 4.3.1. Figure 4.8 shows the hierarchical classification of the

Table 4.2 Inter-object and intra-object mean distances of Fig. 4.7

Object	box (a)	onion (b)
box (a)	12.89	114.90
onion (b)	114.90	13.81

eight objects grouped into three main kinds of objects. The tree outlined by the dendrogram clearly shows groupings of objects based on similar content and suitable similarities between 'families' of objects, e.g., cars were more similar to cans than onions. Thus, three groupings of the objects were found: cars, cans, and vegetables.

Figure 4.9 shows a second set of object images from the COIL-100 database. These objects were recognized and classified following the procedure explained above; the results are shown in Fig. 4.10. The clustering algorithm found meaningful groupings that describe the objects at higher hierarchy levels from the basis extracted in ICA mixtures at the bottom of the hierarchy.

4.5.2 Image Segmentation

The proposed algorithm was applied to segmentation of natural images. The goal was to obtain a meaningful bottom-up structure merging several zones of an image. Figure 4.11 shows an image with nine zones, some of which are clearly different and other which are more or less similar to each other. The total size of the image is 449×512 pixels. In all the experiments of image segmentation the following was done: (i) a set of 1,000 image patches (windows) of 8×8 pixels were taken at random location from each zone, (ii) the normalization, whitening, and dewhitening procedure explained above was applied, (iii) the number of classes of the Mixca algorithm was configured to be the number of zones of the image, (iv) supervised training was used to estimate the ICA parameters for the lowest level of the hierarchy, and (v) a hierarchical representation using the proposed clustering algorithm was obtained.

The dendrogram in Fig. 4.11 shows how the zones are merged from the basis functions. Four segments have been found: sky, cone, structure (roof) and persons, and stairs. These segments are grouped into two broad segments distinguishing the kinds of basis functions that correspond to the part of the image that mainly contains portions of sky, and those zones that correspond to patches where there is a predominant portion of stairs (high frequency). The dendrogram also shows the distances at which the clusters are merged; it can be used as a similarity measure of the zones of the image. The bottom zones are merged at low distances due to the high similarity in borders. Therefore, the hierarchical structures obtained from the zones of the natural image allow for an intuitive interpretation of the scene from different degrees of generalization. This is significant since it can be related with a complex abstraction process.

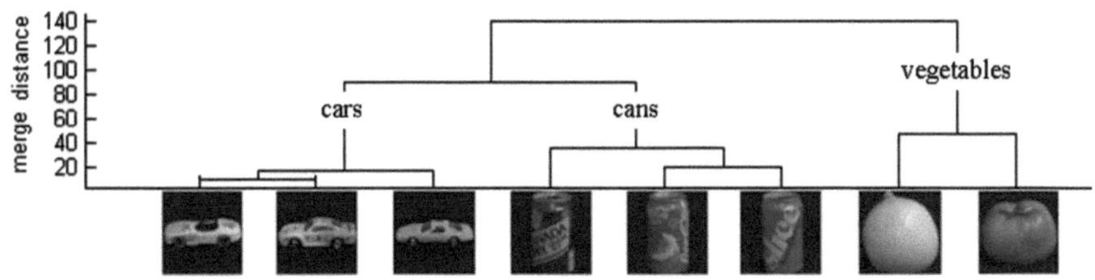

Fig. 4.8 Hierarchical representation of object agglomerative clustering. Three kinds of object 'families' were obtained

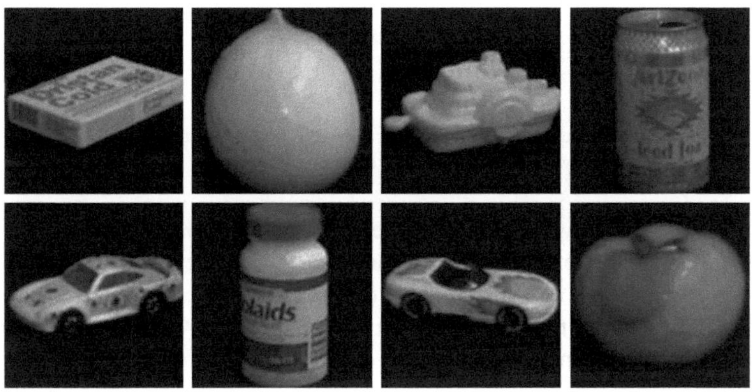

Fig. 4.9 Some object images from the COIL-100 database (before grouping)

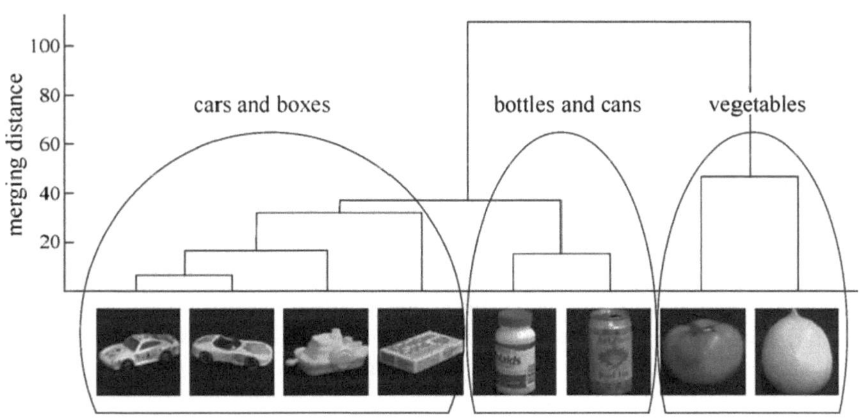

Fig. 4.10 Hierarchical clustering from the objects of Fig. 4.9

Figure 4.12 to an image of 1,344 $\times$ 800 pixels that are divided into 16 zones. The order of zone numbering is left to right at columns and top to down at rows; for instance, in Fig. 4.12 (left) the lowest zone number (1) is located at the top left corner and the highest zone number (16) is at bottom right corner of the image.

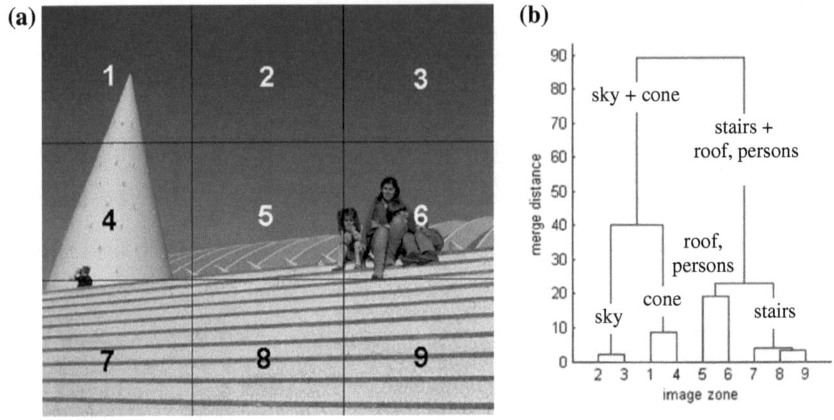

Fig. 4.11 Image segmentation. **a** Image divided in nine zones, **b** hierarchical representation of the zones of the image. Two broad groups of zones are shown

Figure 4.12 (left) shows a mixed image that includes two subimages: a natural image (a frog) and a text image. There are clear differences in the borders of each subimage, which are indicated in the distances at which these subimages are merged at the penultimate level of the hierarchy (Fig. 4.12 (right)). Thus, the segmentation of the image into the two different subimages has been found.

4.6 Conclusions

A method for agglomerative hierarchical clustering based on an underlying ICA mixture model has been proposed. The new algorithm uses the ICAMM parameters estimated at the bottom of the hierarchy to create higher levels by grouping clusters. It is based on the symmetric Kullback–Leibler divergence between the clusters using the ICA parameters assuming non-parametric kernel-based source densities. Different structures of classification can be derived at the different levels of the bottom-up merging. A stopping criterion to estimate an optimum cluster partition was applied using the partition and partition entropy coefficients.

The method has been tested by means of simulations and real data analyses. The simulations showed the capability of the method to generalize from close data densities and to detect outliers. This was compared with the traditional single linkage method that is based on distance between data objects. The proposed methods showed more suitable groupings than the single linkage method (since not only the distance between the data objects is taken into account, but also the kinds of distributions).

The results demonstrated the suitability of the proposed method to process image data. Image content similarity between objects based on ICA basis functions allow an organization of objects in higher levels of abstraction to be learned. In

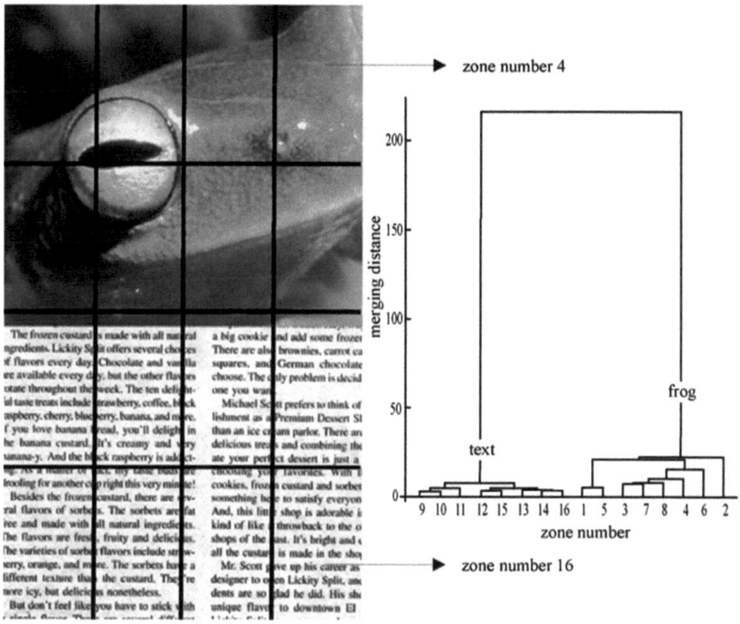

Fig. 4.12 Example of image segmentation—natural and text subimages

this type of organization, the more separated hierarchical levels are, the more different the objects are. The experiments with natural images demonstrated meaningful hierarchy groupings of image zones with similar textures and borders. Thus, image segmentation based on the content of the different zones was obtained. The application of the procedure could be extended to unsupervised or semi-supervised classification of images in order to discover meaningful contents in the hierarchical levels. Thus, applications such as content-based image retrieval in semantic spaces could be attempted.

Finally, note that the supervised-unsupervised scheme of the procedure (to estimate the ICA parameters at the bottom of the hierarchy) would facilitate testing and building the signal database in real-world applications.

References

1. B. Everitt, S. Landau, M. Leese, *Cluster Analysis*, 4th edn. (Arnold, London, 2001)
2. R. Xu, D. Wunsch, Survey of clustering algorithms. IEEE Trans. Neural Netw. **16**(3), 645–678 (2005)
3. D.T. Pham, A.A. Afify, Clustering techniques and their applications in engineering. Proc. Inst. Mech. Eng. Part C J. Mech. Eng. Sci. **221**(11), 1445–1459 (2007)
4. G. Lance, W. Williams, A general theory of classification sorting strategies. 1. Hierarchical systems. Comput. J. **9**(4), 373–380 (1967)

5. A.K. Jain, M.N. Murty, P.J. Flynn, Data clustering: a review, ACM Comput. Surv. **31**(3) (1999)
6. C. Williams, A MCMC approach to hierarchical mixture modelling. Int. Conf. Neural Inf. Process. Sys. NIPS **13**, 680–686 (1999)
7. R.M. Neal, Density modeling and clustering using Dirichlet diffusion trees. Bayesian Stat. **7**, 619–629 (2003)
8. C. Kemp, T.L. Griffiths, S. Stromsten, J.B. Tenenbaum, Semi-supervised learning with trees, Int. Conf. Neural Inf. Process. Sys. NIPS **17** (2003)
9. N. Vasconcelos, A. Lippman, Learning mixture hierarchies. Int. Conf. Neural Inf. Process. Sys. NIPS **12**, 606–612 (1998)
10. A. Stolcke, S. Omohundro, Hidden Markov model induction by Bayesian model merging. Int. Conf. Neural Inf. Process. Sys. **6**, 11–18 (1992)
11. J.D. Banfield, A.E. Raftery, Model-based Gaussian and non-gaussian clustering. Biometrics **43**, 803–821 (1993)
12. S. Vaithyanathan, B. Dom, Model-based hierarchical clustering. Uncertain Artif. Intell. **16**, 599–608 (2000)
13. E. Segal, D. Koller, D. Ormoneit, Probabilistic abstractions hierarchies. Int. Conf. Neural Inf. Process. Sys. NIPS **15**, 913–920 (2001)
14. M.F. Ramoni, P. Sebastiani, I.S. Kohane, Cluster analysis of gene expression dynamics. Natl Acad Sci **99**, 9121–9126 (2003)
15. N. Friedman, Pcluster: probabilistic agglomerative clustering of gene expression profiles. Technical report, vol 80 (Herbew University 2003)
16. K.A. Heller, Z. Ghahramani, Bayesian hierarchical clustering, ACM International Conference Proceeding Series. *Proceedings of the 22nd international conference on Machine learning*, vol 119 (Bonn, Germany, 2005), pp 297–304
17. C.M. Bishop, M.E. Tipping, A hierarchical latent variable model for data visualization. IEEE Trans. Pattern Anal. Mach. Intell. **20**(3), 281–293 (1998)
18. M.E. Tipping, C.M. Bishop, Probabilistic principal component analysis. J. R. Stat. Soc. Series B **61**(3), 611–622 (1999)
19. M.E. Tipping, C.M. Bishop, Mixtures of probabilistic principal component analyzers, Neural Comput. **11**(2), 443–482 (1999)
20. H.J. Park, T.W. Lee, Capturing nonlinear dependencies in natural images using ICA and mixture of Laplacian distribution. Neurocomputing **69**, 1513–1528 (2006)
21. D.J. Mackay, *Information Theory, Inference, and Learning Algorithms* (Cambridge University Press, Cambridge, 2004)
22. F.R. Bach, M.I. Jordan, Beyond independent components: trees and clusters. J. Mach. Learn. Res. **3**, 1205–1233 (2003)
23. A. Hyvärinen, P.O. Hoyer, M. Inki, Topographic independent component analysis. Neural Comput. **13**(7), 1527–1558 (2001)
24. R.S. Raghavan, A method for estimating parameters of K-distributed clutter, IEEE Trans. Aerosp. Electron. Sys. **27**(2), 268–275 (1991)
25. J.C. Bezdek, *Pattern Recognition with Fuzzy Objective Function Algorithms* (Plenum Press, New York, 1981)
26. A.J. Bell, T.J. Sejnowski, The "independent components" of natural scenes are edge filters. Vis. Res. **37**(23), 3327–3338 (1997)
27. J.H. Van Hateren, A. van der Shaaf, Independent component filters of natural images compared with simple cells in primary visual cortex. Proc R Soc Lond B **265**, 359–366 (1998)
28. Y. Matsuda, K. Yamaguchi, Linear multilayer ICA generating hierarchical edge detectors. Neural Comput. **19**(1), 218–230 (2007)
29. T.W. Lee, M.S. Lewicki, T.J. Sejnowski, ICA mixture models for unsupervised classification of non-gaussian classes and automatic context switching in blind signal separation. IEEE Trans. Pattern Anal. Mach. Intell. **22**(10), 1078–1089 (2000)

30. T.S. Lee, D. Mumford, Hierarchical Bayesian inference in the visual cortex. J. Opt. Soc. Am. A **20**(7), 1434–1448 (2003)
31. S.A. Nene, S.K. Nayar, H. Murase, Columbia Object Image Library (COIL-100), Technical report CUCS-006-96, February (1996)
32. A. Hyvärinen, J. Karhunen, E. Oja, *Independent Component Analysis* (Wiley, New York, 2001)

Chapter 5
Application of ICAMM to Impact-Echo Testing

5.1 Introduction

Having information about the condition of a material is an important issue for many industries. This is especially valued if the applied procedure is not time-consuming and is easy to employ in the production line. This is the case of the so-called impact-echo method, which is simply based on making an impact in the material being analyzed. Nevertheless, the impact-echo method is essentially limited to obtaining information about the general status of the specimen. When more detailed information is required (e.g. kind, orientation, and dimension of the defects), other time-consuming and more costly methods, like ultrasonic tomography, are required. In this work, we aim to improve the capability of the impact-echo technique in order to derive more detailed information about the possible defects of the material.

In the impact-echo technique, a material is impacted with a hammer, which produces a response that is sensed by a mono- or multi-sensor system that is located on the surface of the material. Thus, the surface motion resulting from the short-duration mechanical impact can be monitored. We apply a multi-channel configuration with sensors located on different sides of a parallelepiped-shaped material. This configuration allows the microstructure material response to be measured from different planes in order to obtain a more complete examination of the underlying wave propagation phenomenon. The impact-echo signals contain backscattering from grain microstructure as well as information about flaws in the inspected material [1]. This technique has been widely used in applications of concrete structures in civil engineering. Cross-sectional modes in impact-echo signals have been analyzed in elements of different shapes, such as circular and square beams, beams with empty ducts or cement fillings, rectangular columns, post-tensed structures, and tendon ducts. Impact-echo has been used in determining superficial crack depth, evaluation of early-age concrete hardness, evaluation of structural integrity, crack propagation tracing and detection of steel corrosion damage in

A. Salazar, *On Statistical Pattern Recognition in Independent Component Analysis Mixture Modelling*, Springer Theses, DOI: 10.1007/978-3-642-30752-2_5,

concrete reinforcement. A displacement of the fundamental frequency to a lower value is the key to identifying the presence of a crack [1]. The physical phenomenon of impact-echo corresponds to wave propagation in solids. When a disturbance is applied suddenly at a point on the surface of a solid, the disturbance propagates through the solid as three different types of waves: P-wave (normal stress), S-wave (shear stress), and R-wave (surface or Rayleigh) [2]. After a transient period in which the first waves arrive, wave propagation becomes stationary in resonant modes that vary depending on the defects inside the material.

The applications of ICA and BSS are extensive in several areas such as biomedical signal processing, audio signal separation, and image processing [3, 4]. Specifically, there are relatively few references of the use of ICA algorithms in NDT [5, 6]. The main difficulties of the application of BSS to vibration signals were analyzed in [7]. They included the following: scaling and labelling indeterminacies of the sources; the dynamic nature of the mechanical systems, which requires a convolutive mixture of sources to be described; the physical relevance of the source meaning; the determination of the exact number of sources *a priori*; the problem of handling signals that are distributed in time and space; and the requirement of system invertibility. In order to handle these difficulties, in [7], J. Antoni proposed focusing the BSS problem on the separation of vibration signals into contributions of periodic, random stationary and random nonstationary sources.

A method for acoustic emission characterization that is based on ICA and higher-order statistics (HOS) applied to ring-type samples from steel pipes for the oil industry was proposed in [8]. This method allowed low signal-to-noise ratio (SNR) sources that were buried in mechanical non-gaussian noise to be separated by taking advantage of the statistical independence basis of ICA. In addition, ICA has been used to detect vibratory signals from termite activity in wood by separating termite alarm signals generated in wood from known signals [9]. It has also been used to identify embedded transient low-level events (combustion-related noise sources such as combustion, fuel injection, piston lap, and valve operation) in diesel engines [10, 11]. These works show that the separation of sources with small energy levels is possible by using ICA since it is based on the statistical independence of the components and not on the energy associated to each frequency component. Recently, the application of BSS to mechanical signals has been adapted to extract only one signal of interest (or sequentially, more than just one). This approach is referred to as blind signal extraction (BSE) or semi-blind source separation (SBSS) since it exploits *a priori* knowledge about the signal of interest. One example of BSE is the extraction of the mechanical signature of one particular fault in the system for gearbox diagnostics [6].

We developed an application of ICA in the field of impact-echo testing in [12, 13]. In the first approach, the transfer functions between the impact point and the defects in the material were modelled as "sources" for blind source separation. In this work was considered that the sensors located on the material surface measured a convolutive mixture of the contribution of each of the defects. From spectral analysis, the dominant resonance frequencies that vary from homogeneous to defective material were selected. The signal spectral content at the selected frequencies was processed by instantaneous ICA instead of dealing with the whole convolutive problem [14, 15]. The stability of the BSS solution was analyzed using bootstrap resampling [16], obtaining a separability matrix to group the estimated source signals in separable subspaces. The separable estimated source signals were compared with the theoretical response of the material (calculated by transient dynamic analysis through three-dimensional finite element models) for determining the reliability of the defect detection. The results showed that source estimates fit well with the theoretical response of the material. In addition, was found that the number of defects can be estimated by ICA in simulations and experiments with various defective parallelepiped-shape materials of aluminium alloy series 2,000.

This chapter presents the application of ICA mixture modelling to non-destructive testing based on the impact-echo technique. The application consists of discriminating patterns for material quality control from homogeneous and defective materials inspected by impact-echo testing. This problem is modelled as a mixture of independent component analysis (ICA) models, representing a class of defective or homogeneous material by an ICA model whose parameters are learned from the impact-echo signal spectrum. These parameters define a kind of particular signature for the different defects. The proposed procedure is intended to exploit to the maximum the information obtained with the cost efficiency of only a single impact. To illustrate this capability, four levels of classification detail (material condition, kind of defect, defect orientation, and defect dimension) are defined, with the lowest level of detail having up to 12 classes. The results from several 3D finite element models and lab specimens of an aluminium alloy that contain defects of different shapes and sizes in different locations are included. The performance of the classification by ICA mixtures is compared with linear discriminant analysis (LDA) and with multi-layer perceptron (MLP) classification. We demonstrate that the mass spectra from impact-echo testing fit ICAMM, and we also show the feasibility of ICAMM to contribute in NDT applications in Sect. 5.3.

The chapter also includes a section dedicated to describing the procedures both for simulations and lab experiments employed to acquire the impact-echo signals (Sect. 5.2). The final section includes the conclusions and future line of research of this application (Sect. 5.4).

5.2 Impact-Echo Measurements

5.2.1 Simulated Signals

The set of simulated signals came from the full transient dynamic analysis of 100 simulated models. Several studies have demonstrated a good approximation between the theoretical material response calculated by using finite element method (FEM) and the results obtained in impact-echo experiments [1]. The theoretical response of the volumetric wave propagation of impact-echo testing can be modelled as [17],

$$\frac{\partial T_{ij}}{\partial x_j} = \rho_0 \frac{\partial^2 u_i}{\partial t^2} \tag{5.1}$$

$$T_{ij} = c_{ijkl} S_{kl} \tag{5.2}$$

where ρ_0 is the material density; u_i is the length elongation with respect to the starting point in force direction; $\frac{\partial T_{ij}}{\partial x_j}$ is the force variation in the i direction due to deformations in j directions; c_{ijkl} is the elastic constant tensor (Hooke's law); and S_{kl} is the strain or relative volume change under deformation in side l in direction k in a unitary cube that represents a material element.

The Eqs. (5.1) and (5.2) state that the force variation in the direction i due to the side stresses in directions j of the material elementary cube is equal to the mass per volume (density) times the strain acceleration (Newton's third law in tensorial form). Deriving an analytical solution to problems that involve stress wave propagation in delimited solids is very difficult, and this is the reason why the existing bibliography in this field is not very extensive. Thus, these equations are normally solved numerically by FEM [18].

The simulation models of this work consisted of parallelepiped-shaped materials of 0.07 $\times$ 0.05 $\times$ 0.22 m. (width, height and length), which were supported at one third and two thirds of the block length (direction z). Simulated finite models corresponded to one class of homogeneous models and eleven classes of inhomogeneous models. The dynamic response of the material structure (time-varying displacements in the structure) under the action of a transient load was estimated from the transient analysis. The transient load, i.e., the hammer impact excitation, was simulated by applying a force-time history of a half-sine wave with a period of 64 μs as a uniform pressure load on two elements at the centre of the front side of the model.

The elastic material constants for the simulated material (aluminium alloy series 2,000) were: density 2,700 kg/m3; elasticity modulus 69,500 Mpa.; and Poisson's ratio 0.22. This material simulation model (including the specific values for the elasticity constants) was selected to replicate the specimen (aluminium alloy parallelepipeds) where real experiments were performed afterwards. Certainly, a myriad of different materials and forms could be selected which could influence the obtained results. Thus, for example, it has been reported [19] that some accepted conclusions about the application of the impact-echo method might not be true depending on the Poisson ratio value (which is the ratio of transverse contraction strain to longitudinal extension strain in the direction of stretching force). However, we consider that the conclusions derived in this work should be applicable to other kinds and/or forms of material given that different defect patterns could result in different statistical models. Comparing specimens of the same material and rather similar forms would be the only constraint to take into account.

Elements with dimensions of about 0.01 m. were used in the finite element models. This size can accurately capture the frequency response up to 40 kHz. Surface displacement waveforms were taken from the simulation results at 8 nodes in different locations on the specimen surface. This would be equivalent to the signals that could be measured by sensors in a real experiment. The signals consisted of 5000 samples obtained with a simulation step size of 1e-5 s. (sampling frequency of 100 kHz). To compare simulations with experiments, the second derivative of the displacement was calculated to work with accelerations since the sensors available for experiments were mono-axial accelerometers. These accelerations were measured in the normal direction to the plane of the material surface where the sensors were hypothetically to be located.

5.2.2 Experimental Signals

To perform impact-echo measurements on the test blocks, the equipment used in measuring was composed of five main components: an impact hammer, sensors, a data acquisition module, a signal conditioner, and a notebook (see Fig. 5.1).

- Instrumented impacts hammer 084A14 PCB
- Sensor accelerometers PCB model 353B17. Bandwidth: 0.7–20 kHz. Sensitivity 10 mV./g. and a weight of 1.7 gr. They are fixed to the test specimen for measurements.
- ICP signal conditioner model F482A18. Up to 8 channels with $\times 1$, $\times 10$ and $\times 100$ independent gains.
- Data acquisition module 6067E with BNC connectors, which is able to digitalize up to 16 channels with a total maximum sampling frequency of 1.2 M samples per second. It works in single sampling mode or continuous sampling mode and connects to a PC through a fire wire bus.
- Notebook. This controls the data acquisition module and signal storing.

The following parameters were used for measuring: (i) 100 kHz of sampling frequency per channel; (ii) 5,000 samples acquired per channel; (iii) 16 bits of vertical resolution; (iv) 10 dB of conditioning gain per channel; (v) single acquisition mode; (vi) trigger level = 1 V.; and (vii) trigger channel = channel 0.

Figure 5.2 shows the first 2,000 samples of some of the signals collected. Note the half-sine signal of the impact in channel 1 (positions of the sensors are depicted in Fig. 5.3). The waveform of the measured signals depends on several variables, for instance, impact location, shape of the defect, and the relation between the sensor and the defect location. In the case of Fig. 5.2, the material contained a fairly symmetric defect to the longitudinal axis that yielded similar waveforms in sensors 4 and 5 that were located on opposite faces of the material, parallel to the

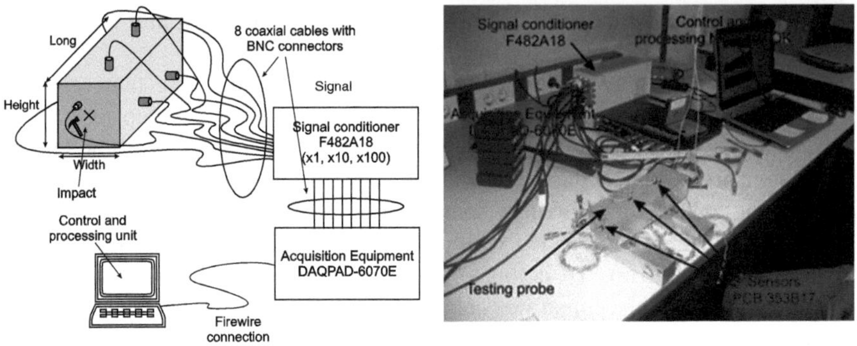

(**a**) Equipment configuration scheme (**b**) Photograph of equipment setup

Fig. 5.1 Equipment setup for impact-echo experiments **a** Equipment configuration scheme **b** Photograph of equipment setup

longitudinal axis. This can also be observed for the sensors 3 and 6, and sensors 7 and 8.

5.3 ICAMM for Learning Material Defect Patterns

This section includes an ICA mixture model applied to the impact-echo signals. There are no previous references to the use of ICAMM in NDT applications. This approach is based on a new model for the joint probability density of features that are extracted from the impact-echo signals as mixtures of ICA models. Hence, an optimum classification method that is based on ICA mixture modelling (ICAMM) is implemented. This study demonstrates that the proposed method is able to distinguish different kinds of defects such as cracks, holes, and combinations of cracks and holes. A low-cost operating NDT procedure is included, and its results are presented for 3D finite element simulations and lab specimens. Note that deriving detailed information from a simple impact is a difficult field of application. Therefore, any results that show some capability of discerning different defect characteristics have great potential interest in material diagnoses.

In Sect. 5.3, we demonstrate that ICA can be used to separate information of material defects in impact-echo testing [13]. In this section, ICAMM is approached from a physical model that is based on dividing the wave path propagation into two parts: impact to point flaws, and point flaws to sensors. It is assumed that the set of point flaws builds defective areas with different geometries, such as cracks (small parallelepipeds), holes (cylinders), and multiple defects (a combination of cracks and holes). Depending on the kind of defective area, the spectrum measured by the sensors changes, which allows the kind of defect condition of the material to be discerned. This study demonstrates that the spectrum of different kinds of defective materials can fit into different ICA models.

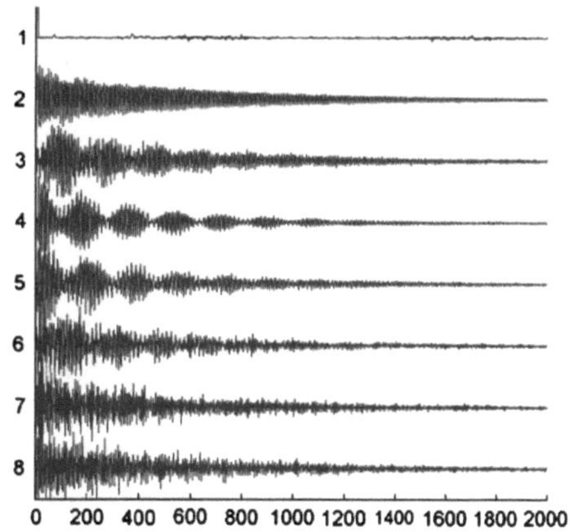

Fig. 5.2 Signals measured in an impact-echo experiment

The quality control problem approached herein consisted of classifying several solid materials that had a parallelepiped shape in homogeneous and different defective classes. The proposed ICA mixture model was tested using an extensive set of impact-echo signals that were extracted from 3D finite element models and lab specimens of a series 2,000 aluminium alloy. This was done because in order to perform controlled and extensive experiments. The models and specimens were classified into four levels (from general to detailed classes) with the lowest level of detail having up to 12 classes. The four classification levels were the following: (i) *Material condition*: homogeneous, one defect, multiple defects; (ii) *Kind of defect*: homogeneous, hole, crack, multiple defects (iii) *Defect orientation*: homogeneous, hole in X axis or Y axis, crack in XY, ZY, or XZ planes, multiple defects; and (iv) *Defect dimension*: homogeneous, passing through or half-passing through holes and cracks of classification level iii, multiple defects.

Figure 5.3 shows outline examples of six of the classes where the defective materials were classified. The other classes corresponded to the homogeneous class and half-passing through versions of the one-defect classes of Fig. 5.3. Note that the multichannel impact-echo system with the location of the eight accelerometers $a1 \cdots a8$ that were used as sensors is depicted for one of the material classes. The data acquisition system has eight channels and one of them (channel $a1$) is used to measure the impact signal. Holes are ϕ=1 cm., and cracks are vertical and extend from the left side to the bulk centre and are 5 mm. wide.

The defects were located at different positions within the pieces, and, in the case of multiple-defect materials, the numbers of defects varied from 2 to 3. The set of simulated and experimental signals were classified using the ICAMM-based classifier Mixca [20], LDA, and MLP 0. The results show that the impact-echo mass spectra fit the ICA mixture modelling. Thus, the theoretical and experimental results of this section provide a basis to undertake experiments at the industrial level.

Although the experiments are focused on specimens of aluminium alloy, there are many potential applications with other kinds and sizes of materials. For instance, in [21], we reported experimental work in the diagnostic of block-sized

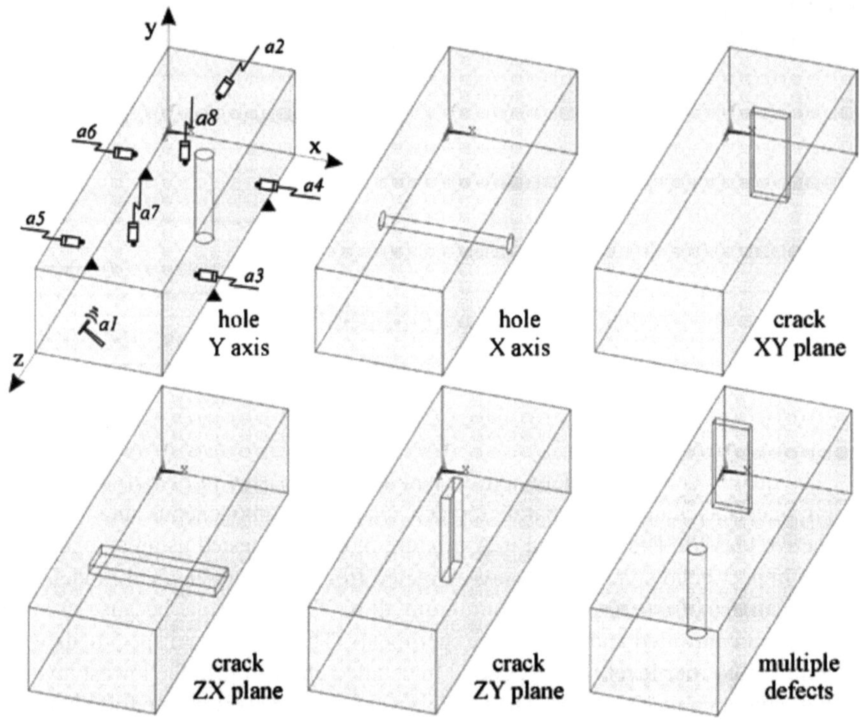

Fig. 5.3 Examples of the different patterns of defective materials. The multichannel setup of the impact-echo inspection is displayed

marble stones (5–9 m^3, 6–10 tons of weight). This study combined single-channel impact-echo testing and ultrasound for general classification of the status of marble rocks into either sound or unsound material. The marble rock defects feature a range of different shapes and sizes with volumes up to the order of cm^3. Thus, an acoustic-wave NDT system is appropriate for the application of general marble diagnosis with capabilities to determine material defective zones and to classify the general status of the block, depending on a given detail level of defect types.

5.3.1 ICA Mixture Statement of the Problem

First of all, note that the aim of this section is to find an underlying ICA mixture model from physical reasoning. This will contribute to a better understanding of the suitability of ICAMM to the impact-echo problem, as well as contribute to a better interpretation of the obtained results. A detailed description of the proposed procedure and the criteria for fitting the different parameters involved is included in the experimental section of this approach.

The degrees of freedom afforded by mixtures of ICA suggest that it is a good candidate for a broad range of problems. As was commented in Chap. 1, there are different perspectives to undertake the ICA mixture modelling of the physical phenomenon under analysis. From the perspective of "most physical" interpretation, this section includes a modelling of the impact-echo data as an ICA mixture. It is clearly important to have as much knowledge as possible about the underlying physical phenomenon, in order to better interpret the results for the general performance of the method.

We proposed an ICA model for the impact-echo problem in Sect. 5.3. This model considered the transfer functions between the impact location and the point defects that are spread in a material bulk as "sources" for blind source separation. In this work, we formulate a model based on ICAMM that takes into account the resonance phenomenon involved in the impact-echo method. The proposed model extends to defects with different shapes, such as cracks or holes, and defines the quality condition determination of homogeneous and defective materials as an ICA mixture problem.

In ICA mixture modelling, it is assumed that feature (observation) vectors $\mathbf{x}_k$ corresponding to a given class C_k $(k = 1\ldots K)$are the result of applying a linear transformation defined by matrix$\mathbf{A}_k$ to a (source) vector $\mathbf{s}_k$, whose elements are independent random variables, plus a bias vector $\mathbf{b}_k$, i.e.,

$$\mathbf{x}_k = \mathbf{A}_k\mathbf{s}_k + \mathbf{b}_k \quad k = 1,\ldots,K \tag{5.8}$$

Let us find the ICA mixture model for the impact-echo problem. The impact-echo signals can be considered as a convolutive mixture of the input signal and the defect signals inside the material, as shown in Fig. 5.4.

In Fig. 5.4, there is one attack point that generates the wave $r_0(n) = p(n)$; Finternal focuses (point flaws) that generate the waves $f_j(n) \quad j = 1,\ldots,F$; and N sensors that measure the waves $v_i(n) \quad i = 1,\ldots,N$. To simplify, we consider the impact as another focus; thus,$f_j(n) \quad j = 0,\ldots,F$, with$f_0(n) = r_0(n)$.

We assume that the impact-echo overall scenario can be modelled as a multiple-input-multiple-output linear time invariant system (MIMO-LTI) [22]. This implies that the complex spectrum at sensor i (Fourier transform of $v_i(n)$) will be the sum of all the contributions due to focus$j = 0,\ldots,F$. Moreover, the contribution of every individual focus j will be the product of the complex spectrum at focus j by the frequency response of the path between focus j and sensor i. In practice, the discrete Fourier transform (DFT) is to be used to allow numerical computation of the complex spectrum. Considering that the length of the recorded signal is long enough to capture the transients, and so to overcome the time-overlapping effect of the DFT, the MIMO-LTI model may be expressed in an algebraic form by means of properly defined vectors and matrices.

Let us call $\mathbf{V}_i$ a vector formed by the samples obtained from the computation of the DFT of $v_i(n)$. Hence, $\mathbf{V}_i$ is a vector representation of the complex spectrum of $v_i(n)$. Considering the MIMO-LTI model, the total complex spectrum $\mathbf{V}_i$ will be

Fig. 5.4 Wave propagation scheme proposed for an inspection by impact-echo using four sensors. The path between the point flaws and the sensors is depicted only for a few focuses

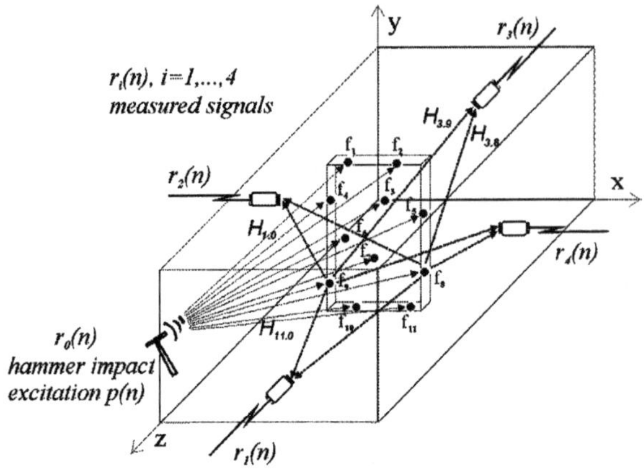

(a) Material with 11 internal focuses due to point flaws that build a crack-shaped-like defect that is oriented in the plane xy

the sum of $F+1$ contributions due to the $F+1$ focuses that are present. Moreover, if we call vector $\mathbf{V}_{ij}$ the contribution to $\mathbf{V}_i$, due to focus j, we can write

$$\mathbf{V}_i = \sum_{j=0}^{F} \mathbf{V}_{ij} = \sum_{j=0}^{F} \mathbf{B}_{ij} \cdot \mathbf{F}_j, \tag{5.9}$$

where $\mathbf{B}_{ij}$ is a diagonal matrix that has the vector formed by the samples of the frequency response between focus j and sensor i at its main diagonal, and where $\mathbf{F}_j$ is a vector formed by the samples of the complex spectrum of $f_j(n)$. On the other hand, the focus signals $\mathbf{F}_j$ $j = 1 \ldots F$ that model the defect response can be expressed in terms of the impact excitation $\mathbf{F}_0$

$$\mathbf{F}_j = \mathbf{M}_{0j} \cdot \mathbf{F}_0 \quad j = 1 \ldots F, \tag{5.10}$$

where $\mathbf{M}_{0j}$ is a diagonal matrix that has the vector formed by the samples of the frequency response between the impact point and focus j at its main diagonal. Combining Eqs. (5.9) and (5.10) and defining $\mathbf{M}_{00} = \mathbf{I}$, we can write

$$\mathbf{V}_i = \sum_{j=0}^{F} \mathbf{H}_{ij} \cdot \mathbf{F}_0 \qquad \mathbf{H}_{ij} = \mathbf{B}_{ij}\mathbf{M}_{0j}, \tag{5.11}$$

where $\mathbf{H}_{ij}$ is a diagonal matrix that has the vector formed by the samples of the frequency response modelling the path between the impact point and focus j, and between the focus j and sensor i at its main diagonal.

The dimensionality of all the vectors and matrices involved in Eq. (5.11) depends on the size of the DFT used. For example, in the application of this work, a size of 1024 is considered since it is large enough to both capture the transients of the signals at the used frequency sampling and to have adequate frequency resolution. Such long vectors imply a great computational burden and convey a lot of redundant information. That is why PCA is often applied to extract, from the original feature vectors, uncorrelated components that describe most of the variance. PCA has been employed to solve problems of overdetermined BSS (the case where more sensors than source signals are available) [23] and to improve the classification accuracy (see for instance [24, 25]). PCA is equivalent to expressing model in Eq. (5.11) in a subspace of properly selected dimension where most of the original variance is explained

$$\mathbf{PV}_i = \sum_{j=0}^{F} \mathbf{PV}_{ij} = \sum_{j=0}^{F} \mathbf{PH}_{ij}\mathbf{P}^T \cdot \mathbf{PF}_0 \Leftrightarrow \mathbf{V}_i^{(PCA)} = \sum_{j=0}^{F} \mathbf{H}_{ij}^{(PCA)}\mathbf{F}_0^{(PCA)}, \qquad (5.12)$$

where $\mathbf{P}$ is a projection unitary matrix ($\mathbf{P}^T\mathbf{P} = 1$) provided by PCA that allows the observation vectors to be ordered by their powers. PCA finds a rotated orthogonal system such that the elements of the original vectors in the new coordinates become uncorrelated, so the redundancy induced by correlation is removed [3]. Dimension of $\mathbf{V}_i^{(PCA)}$depends on the variance to described on it with respect to the variance on the original vector. For example, in the experiments, we reduce the number of features from 512 (the first half of the DFT) to only 20, thus retaining 95 % of the total variance of the data.

In order to obtain one only feature vector corresponding to the complete MIMO-LTI model of the multichannel impact-echo setup, let us form one vector from all vectors $\mathbf{V}_i^{(PCA)}$ $i = 1\ldots N$

$$\begin{bmatrix} \mathbf{V}_1^{(PCA)} \\ \vdots \\ \mathbf{V}_N^{(PCA)} \end{bmatrix} = \begin{bmatrix} \sum_{j=0}^{F} \mathbf{H}_{1j}^{(PCA)}\mathbf{F}_0^{(PCA)} \\ \vdots \\ \sum_{j=0}^{F} \mathbf{H}_{Nj}^{(PCA)}\mathbf{F}_0^{(PCA)} \end{bmatrix} \Leftrightarrow \mathbf{V}^{(PCA)}$$
$$= \sum_{j=0}^{F} \begin{bmatrix} \mathbf{H}_{1j}^{(PCA)} \\ \vdots \\ \mathbf{H}_{Nj}^{(PCA)} \end{bmatrix} \cdot \mathbf{F}_0^{(PCA)} = \sum_{j=0}^{F} \mathbf{H}_j^{(PCA)} \cdot \mathbf{F}_0^{(PCA)} \qquad (5.13)$$

Vector $\mathbf{V}^{(PCA)}$ implies a dimension increase by a factor N in comparison with vectors $\mathbf{V}_i^{(PCA)}$. Moreover, depending on the sensor spatial distribution, some correlation may exist between components of $\mathbf{V}^{(PCA)}$ corresponding to different sensors. Hence, a new PCA projection matrix $\mathbf{P}_1$ should be applied in a similar manner to Eq. (5.12), resulting in:

$$\mathbf{P}_1\mathbf{V}^{(PCA)} = \sum_{j=0}^{F} \mathbf{P}_1\mathbf{H}_j^{(PCA)}\mathbf{P}_1^T \cdot \mathbf{P}_1\mathbf{F}_0^{(PCA)} \Leftrightarrow \quad \mathbf{V}^{((PCA))} = \sum_{j=0}^{F} \mathbf{H}_j^{((PCA))}\mathbf{F}_0^{((PCA))}. \qquad (5.14)$$

In the experiments, the dimension of $\mathbf{V}^{(PCA)}$ is 140 (20 components of $\mathbf{V}_n^{((PCA))}$ by $N = 7$ sensors), and it is reduced to only 50 components in $\mathbf{V}^{((PCA))}$, thus retaining 92 % of the total variance.

Now, in order to account for the variability of the impact generation (different strength, different locations,..), we assume that $\mathbf{F}_0^{((PCA))}$ can be expressed as a mean excitation $\mathbf{m}$ plus a random variation $\mathbf{s}$. Substituting in equation (5.14)

$$\mathbf{V}^{((PCA))} = \sum_{j=0}^{F} \mathbf{H}_j^{((PCA))}(\mathbf{m} + \mathbf{s}) = \mathbf{A}\mathbf{s} + \mathbf{b} \; ; \; \mathbf{A} = \sum_{j=0}^{F} \mathbf{H}_j^{((PCA))} \; ; \; \mathbf{b} = \sum_{j=0}^{F} \mathbf{H}_j^{((PCA))}\mathbf{m} \tag{5.15}$$

Equation (5.15) demonstrates the suitability of an ICA model for the multichannel impact-echo scenario. In a given specimen, an ICA model can be applied to estimate $\mathbf{A}$ and $\mathbf{b}$ from a set of training feature vectors $\{\mathbf{V}_n^{((PCA))}\}$obtained by repeated impacts on the material. Note that linear transformation $\mathbf{A}$ depends on the transfer functions between the focus and the sensors and between the excitation point and the focus, whereas the bias term $\mathbf{b}$ additionally depends on the mean impact excitation. This indicates that, in principle, a different ICA model should be required for every specific defect (defective zone with particular geometry), every specific deployment of the sensors, and every specific impact location. Thus, we can formulate the problem of classification of materials with different quality conditions, which are inspected by impact-echo in the ICAMM framework. Equation (5.15) can be formulated to the case of a given material class C_k $(k = 1 \ldots K)$, considering a different set of parameters $\mathbf{A}_{(k)}$, $\mathbf{s}_{(k)}$ for every class of material. The following ICAMM expression can be written,

$$\mathbf{V}_{(k)}^{((PCA))} = \mathbf{A}_{(k)}\mathbf{s}_{(k)} + \mathbf{b}_{(k)} \; ; \quad \mathbf{A}_{(k)} = \sum_{j=0}^{F} \mathbf{H}_{j(k)}^{((PCA))} \; ; \; \mathbf{b}_{(k)} = \sum_{j=0}^{F} \mathbf{H}_{j(k)}^{((PCA))}\mathbf{m}_{(k)}, \tag{5.16}$$

where,

$\mathbf{V}_{(k)}^{((PCA))}$ compressed spectra of the multichannel impact-echo setup for the defective material class k

$\mathbf{A}_{(k)}$ mixture matrix for the defective material class k

$\mathbf{s}_{(k)}$ compressed spectra from the focuses $f_j, j = 0, \ldots F$ for the defective material class k

Finally, let us express Eq. (5.11) considering separately the magnitude and the phase term of the complex numbers, i.e.,

$$\mathbf{V}_i = \sum_{j=0}^{F} \Phi_{ij}\left|\mathbf{H}_{ij}\right| \cdot \Phi_0\left|\mathbf{F}_0\right|, \tag{5.17}$$

where Φ_{ij} and Φ_0 are diagonal matrices that have the corresponding exponential phase terms at their main diagonal, and where $\left|\mathbf{H}_{ij}\right|$ and $\left|\mathbf{F}_0\right|$ are the matrix and vector, respectively, including the magnitude of the elements of $\mathbf{H}_{ij}$ and $\mathbf{F}_0$ (remember that $\mathbf{H}_{ij}$ is a diagonal matrix). Now note that when the shortest wavelength involved is much greater than the defect size, the delays due to propagation between the impact point and the different focus locations may be considered equal for all focuses. The same can be assumed for the delays due to propagation between the focus location and every specific sensor. This implies that $\Phi_{ij} \simeq \Phi_i$ will be the same for $j = 0, \ldots, F$. Therefore, we can write

$$\mathbf{V}_i \simeq \Phi_{ij}\Phi_0 \sum_{j=0}^{F} \left|\mathbf{H}_{ij}\right| \cdot \left|\mathbf{F}_0\right| \quad \Rightarrow \quad \left|\mathbf{V}_i\right| \simeq \sum_{j=0}^{F} \left|\mathbf{H}_{ij}\right| \cdot \left|\mathbf{F}_0\right| \tag{5.18}$$

Then a derivation similar to the one we made from Eqs. (5.11–5.15) working with the complex spectrum could now be made from Eq. (5.18) considering only the magnitude spectrum. Using the magnitude spectrum makes the application of ICA simpler, as most ICA algorithms are appropriate only for the real case. The shortest wavelength is given by $\lambda_{\min} = c/B$, where c is the speed of propagation of sound in the material under analysis and B is the bandwidth of the impact signal. In the experiments, $c = 4800$ m/s and $B = 20$ KHz, so that $\lambda_{\min} = 24$ cm, and hence it seems reasonable to work with the magnitude spectrum.

The model explained above was tested in simulations and experiments of impact-echo testing. The procedure to acquire the signals was the explained in Sect. 5.2. The number of available records for classifying was 2,100 and 2,030 for simulations and experiments, respectively. These records correspond to 100 3D FEM models and 92 specimens of aluminium alloy series 2,000. The sample of simulation records was obtained by adding 21 replicates with spherical Gaussian noise to the 100 original records. For each tested specimen, around 22 repetitions of the impact-echo experiment were performed for a total of 2,030 impact-echo records. The number of lab tested specimens was 92 (parallelepiped-shaped $7 \times 5 \times 22$ cm. bulks of aluminium alloy series 2,000).

5.3.2 Classification Procedure

The error in the reconstruction of a particular data vector is shown in Fig. 5.5, corresponding to the spectrum estimated from a signal acquired in channel 4 of a piece with a crack plane XZ defect. Three plots are displayed: original data, reconstructed data, and residual data (difference between original and

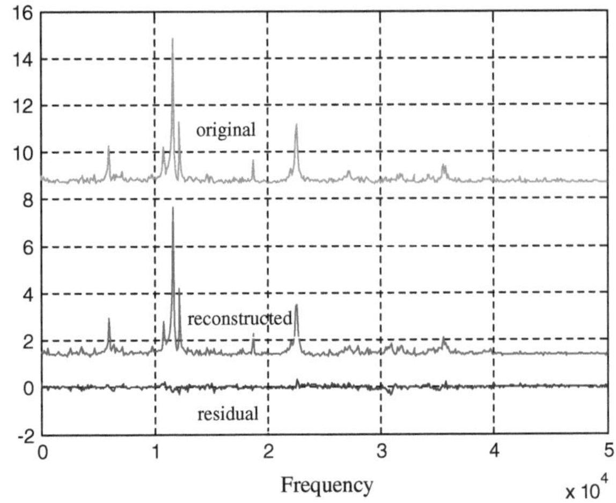

Fig. 5.5 Reconstruction of a spectrum from a signal acquired in channel 4 of a piece with a XZ plane crack defect

reconstructed data). The reconstructed spectrum was obtained using the first 20 principal components (explained variance ≥ 95 %). It can be seen that the reconstructed data is smoother than the original data and the residual data do not contain significant features.

One problem that arises from the reduction of dimensionality is the possible loss in the reconstructed data of discriminator weak features that are present in a small portion of the data. Those features exhibit low correlation with the rest of the data, and PCA will allocate them to the least significant components. However, those rare features could be spurious due to instrumental issues, and so on, and they have to be filtered before the classification. The impact-echo data of this application did not show those kinds of features, as Fig. 5.5 shows.

In order to obtain a mass spectra version for the multichannel setup of the impact-echo experiments, the vectors of reduced dimensionality $\mathbf{V}_i^{(PCA)}$ $i = 1...N$ of each experiment were arranged into a single vector $\mathbf{V}^{(PCA)}$ with dimensionality increasing by a factor of N, i.e., dimensionality $= 7$ x $20 = 140$ (see Equ. (5.13)). PCA was applied to $\mathbf{V}^{(PCA)}$ to obtain a version of reduced dimensionality $\mathbf{V}^{((PCA))}$ of the data vectors in Eq. (5.14). The number of components retained was estimated at 50, reducing the dimensionality from 140 to 50, obtaining a reconstructed variance ≥ 92 % of the variance of the $\mathbf{V}^{(PCA)}$ data vectors. Thus, the final dimensions of the input data matrices for the classification stage were (2,100 $\times$ 50) and (2,030 x 50) for simulations and experiments, respectively. A rationale of the PCA application and selection of a certain number of components is included in [26].

Figure 5.6 shows an outline of the steps followed above in order to obtain the data vectors for classification and the steps of the classification stage to obtain the results. The dimensionality of the data vectors was 50, corresponding to a compressed mass spectra data set from the multichannel impact-echo simulations and experiments. The classifiers employed in the classification stage were: LDA, MLP, and Mixca. The technique applied to select the training and testing samples was leave-one-out with supervised training. All the records, including repetitions or replicas, corresponding to a piece to be tested were taken out of the training stage of that piece.

The first step of the classification stage consisted of determining the capacity of the features from the data vectors to discern between the different classes of the four levels of classification (i-material condition, ii-kind of defect, iii-defect orientation, and iv-defect dimension). Thus, a feature selection step was made using the classification results of LDA. It consisted of trying several performances of classification using different data subsets from the original 50-dimension data set. The data subsets were matrices of varying dimension $M \times d, d = 1\ldots50$ for each performance. All the data set vectors were used, but the number of spectral features employed in the classification varied for each performance, increasing from 1 to 50. LDA was applied using the linear, quadratic, and Mahalanobis distances; maximum classification accuracy, was obtained by the quadratic distance.

The feature selection step using the results of LDA classifications with quadratic distance for the experiment data set is shown in Fig. 5.7. The curves describe the classification error as a function of the number of features employed. There is an optimum number of features where the maximum classification accuracy is obtained. Note that an increase in the number of features progressively improves the classification results up to a point where adding more features makes the classification accuracy worse. The optimum number of features varied depending on the classification level; it was 10, 15, 18, and 25, respectively, for classification levels i, ii, iii, and iv.

The features corresponding to the maximum classification accuracy obtained by LDA in the feature selection stage were selected as the input for the classification with MLP and Mixca. LDA and MLP are well-known methods for classification, whereas Mixca has only recently been reported [20]. MLP was applied using a validation stage, one hidden layer with a tuned number of neurons, and a resilient backpropagation learning algorithm. Mixca was applied considering the definition provided in Sect. 5.3.1 of an ICA mixture model for the material quality determination using impact-echo testing, i.e., the data provided for a class of material can be modelled by the parameters of one ICA.

Let us highlight some aspects of the Mixca algorithm explained in Chap. 3. The observation vectors $\mathbf{x}_k$ are modelled as the result of applying a linear transformation $\mathbf{A}_k$ to a vector $\mathbf{s}_k$ (sources), whose elements are independent random variables, plus a bias vector $\mathbf{b}_k$; thus, $\mathbf{s}_k = \mathbf{A}_k^{-1}(\mathbf{x}_k - \mathbf{b}_k)$. Let us call $\mathbf{A}_k^{-1} = \mathbf{W}_k$. The algorithm is based on maximizing the data likelihood $L(\mathbf{X}/\mathbf{\Psi}) = \log p(\mathbf{X}/\mathbf{\Psi}) = \sum_{n=1}^{N} \log p(\mathbf{x}^{(n)}/\mathbf{\Psi})$ where $\mathbf{\Psi}$ is a compact notation for all the unknown parameters $\mathbf{W}_k$, $\mathbf{b}_k$ for all the classes $C_k = (k = 1\ldots K)$. It is considered that, in a mixture model, the probability of every available feature vector can be separated into the contributions that are due to each class.

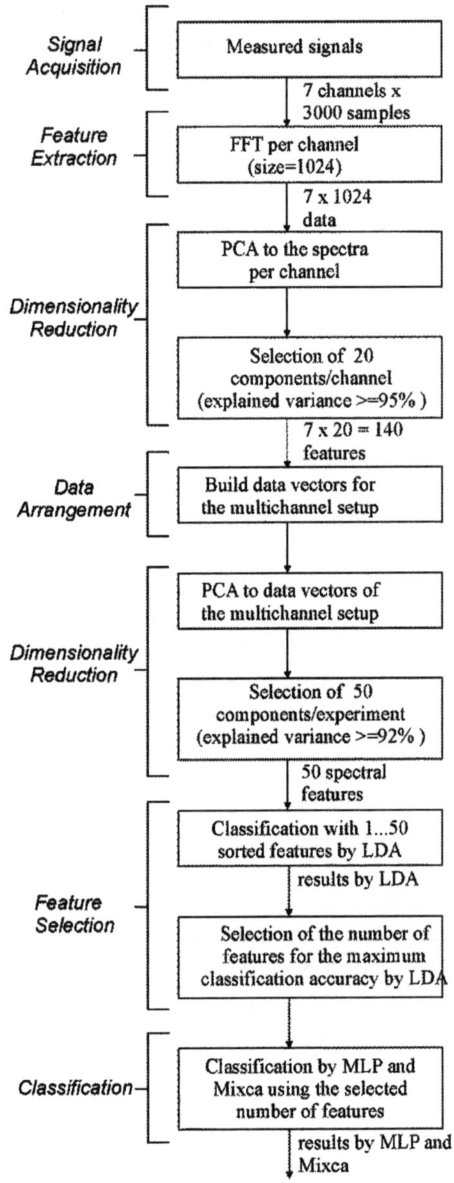

Fig. 5.6 Outline of the classification procedure

We used two different algorithms for the estimation of the $\boldsymbol{\Psi}$ ICA parameters for each class in the Mixca procedure. The first algorithm was JADE (Mixca-JADE) [27], and the second algorithm was the proposed Gaussian kernel-based (non-parametric) approach included in Mixca (see Chap. 3) [20]. This second algorithm uses the expression $p\left(s_{km}^{(n)}\right) = a \cdot \sum_{n' \neq n} e^{\frac{1}{2}\left(\frac{s_{km}^{(n)} - s_{km}^{(n')}}{h}\right)^2}$ $m = 1 \ldots M$ $k =$

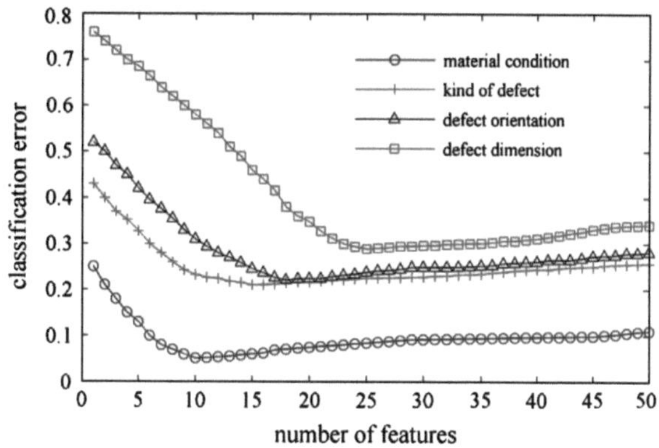

Fig. 5.7 Results of the feature selection stage using LDA classification

$1\ldots K$ to estimate the source densities. The parameters applied in the classifications by Mixca were the following: the kernel parameters were normalization constant a=1; and the constant for defining the degree of smoothing was estimated as $h = 1.06\sigma N^{-1/5}$, which is the normal reference rule using a Gaussian kernel [28]. The learning parameter α of the gradient technique used in Mixca was selected in a rather experimental manner (α = 5e-05). However, some guidance may help in reducing the search for α based on known conclusions about the learning of ICA algorithms [29]. With regard to ICAMM, a discussion is included in [20] which may help to understand the problems of convergence. The likelihood threshold to stop the iterations was 1e-05.

5.3.3 Patterns Detected in ICA Mixtures

Figure 5.8 shows the spectra extracted from the 100 simulated models at channels $a2$ and $a3$ of Fig. 5.3. In the vertical axes, the lower position spectra correspond to homogeneous pieces, and the higher position spectra correspond to multiple-defect pieces. The wave propagation phenomena (depending on the defects) produce frequency shifting, appearance of new tones, and also enlargement of the tones, which can all be clearly observed. In addition, these phenomena are different for different models. It was assumed for the proposed classification that the shifting spectra depend more on the orientation (X or Y axes, and XY, ZY, or XZ planes) and the dimension (passing or not passing through defect) of the defects than on the defect location.

Figure 5.9 shows a scatterplot for the compressed mass spectra of the multi-channel impact-echo setup from the experiments in the kind of defect level (4 classes), The data are drawn in a 3D space defined for the first three spectra features of the data set vectors. Figure 5.9 has been rotated to highlight a point of view where the data for the different classes are discernible; however, there are zones where the data are packed together, which make the separation of samples

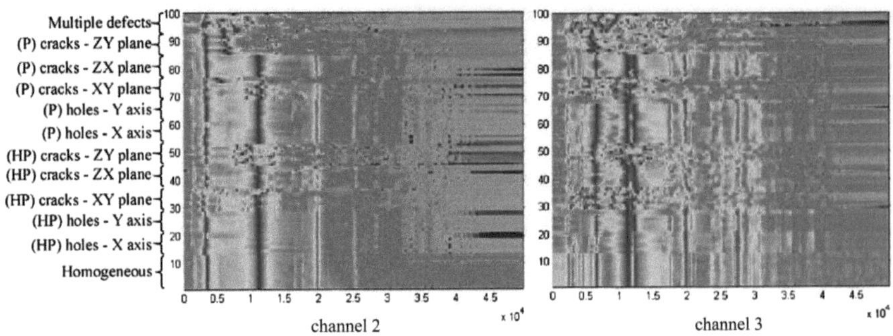

Fig. 5.8 Magnitude spectrum of the impact-echo signals for the simulated models. *Abbreviations P* passing through, *HP* half-passing through channel 2 channel 3

belonging to the four different classes difficult. In addition, not all the data for the multiple-defect class are depicted in Fig. 5.9, since they are very sparse in the classification space. Homogeneous specimen data are concentrated in a small spherical region which makes their classification easy. The crack and hole data are in a wide area and describe straight-like trajectories; they are joined together with some of the multiple-defect data. In spite of this, for this classification level (Level ii- kind of defect), the Mixca algorithm was able to find a proper solution, which reached an accuracy of 88.9%. Thus, the impact-echo signals of each model were fitted to different ICA models whose parameters were estimated by Mixca.

Figure 5.10 shows a set of ICAMM parameters obtained during training for the impact-echo experiments using 15 spectra features in the kind of defect level of classification. The estimated mixing matrices (represented in grey scale) and the sources with their distributions and kurtosis values for the four classes of materials are shown. The ICA parameters estimated for each class showed different sources with non-gaussian distributions. The differences in these parameters among the classes clearly show the suitability of the ICAMM model for classifying different kinds of defective materials inspected by impact-echo.

The estimated sources represent linear combinations of the spectrum elements produced by the defects that activate different resonant modes of the material. In this level of classification, the pattern of the defects was detected independently of their orientation and dimension. These patterns are related to the number of point flaws that build the defects and the spatial relationship between the flaws. In defective materials, the propagated waves have to surround the defects; their energy decreases and multiple reflections and diffraction with the defect borders are produced. The patterns of the displacement waveforms are affected by the shape of the defects [30] building a kind of signature of the defect. This signature is distinguishable in the parameters estimated by Mixca since the mixing matrix is different for every class and there are particular densities of the sources that are recovered only for a specific class.

The results obtained have confirmed the theoretical development included in Sect. 5.3.1 where it was demonstrated that an ICA model can be applied for every

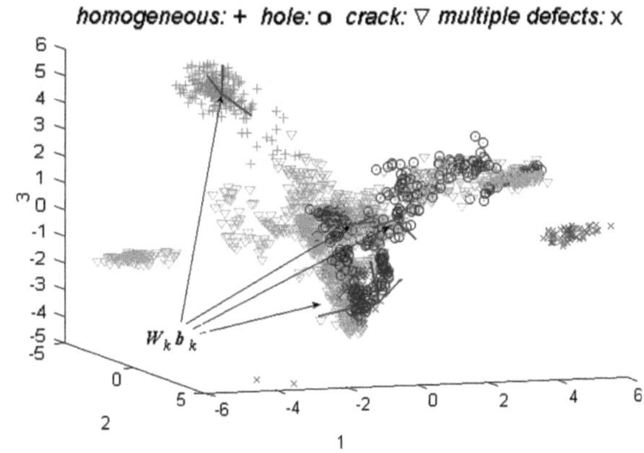

Fig. 5.9 Scatterplot of the experimental data and estimated parameters of the ICA mixture for Level ii of classification: kind of defect. The classification space corresponds to the first three spectra features of the data set vectors

specific defect, every specific deployment of the sensors, and every specific impact location. Since the deployment of sensors and impact location have been kept constant, we have considered grouping different defects having a common characteristic (like orientation, size and/or shape) in the same ICA class of an overall ICAMM model. Hence, it is assumed that by keeping one or more of the defect characteristics constant, a common ICA model could be estimated, because **A** and **s** would not vary very much for the different defects of a class. This assumption is required to define categories that describe, in more or less detail, the inner state of the specimen from the multichannel impact-echo signals.

5.3.4 Results

In this section, we compare the results of Mixca with MLP and LDA. Figure 5.11 shows a summary of the classification results for the classifiers LDA, MLP and Mixca using JADE (Mixca-JADE) and the non-parametric kernel-based density estimation (Mixca) for the ICA parameter updating, for both simulations and experiments (the accuracy is the mean percentage of success in classification). In general, the curves show that the classification accuracy decreases when the number of classes increases in more detailed levels of classification. For experiments, the best results were obtained by non-parametric Mixca; for simulations, the best results were obtained by LDA and non-parametric Mixca.

The experiment results were better than simulation results since the number of specimens and models and the implementation to obtain the signals were different. The experiments of impact-echo involved some randomness in their execution since the force injected in the impact excitation and the positions of the sensors (which can vary from piece to piece) are manually controlled. These variables yield repetitions of the experiments with their corresponding signal spectra. These spectra separated class regions better than the Gaussian noise that was used to obtain replicates of the simulated model signals (since data densities are

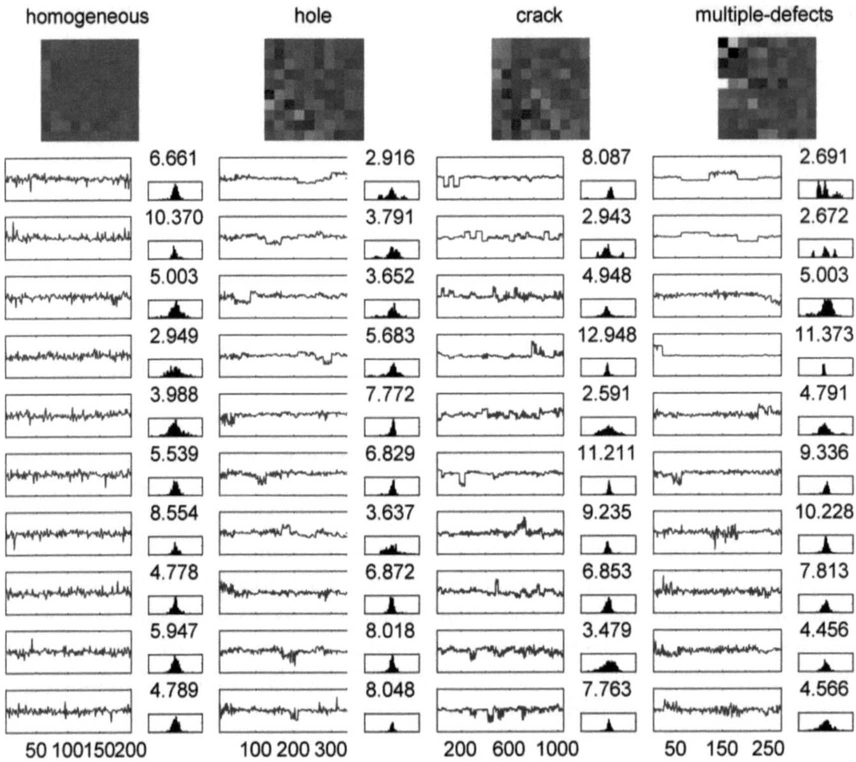

Fig. 5.10 Mixing matrices and sources estimated for four different kinds of defective materials tested in impact-echo experiments. The kurtosis values are displayed for the source densities

smoothed). The results show that ICA-based classifiers perform better in the case of data with implicit non-gaussianity of the sources compared with the other tested classifiers. In addition, the non-parametric estimation of the source densities allowed more accurate data modelling since no restriction on the data model was assumed.

Table 5.1 shows the results obtained by non-parametric Mixca in the experiments for the different levels of classification. The percentage of success decreases steadily for higher grades of knowledge about the material defects. Note that, at the most complex level of classification (12 classes at the defect dimension level), a classification accuracy of 78.1 % is obtained. This may or may not be appropriate depending on the application; however, in any case it is an interesting percentage if we take into account the simplicity of the procedure (one only impact) and the number of classes discerned. The homogeneous specimens are always well classified in all levels; the multiple-defect specimens are the worst classified at higher levels; and the hole specimens show the most cases of misclassification at lower levels.

Table 5.1 Classification accuracy for experiments obtained by non-parametric Mixca through the different levels of the material quality control problem

	% general	*Levels of quality of material*												
Material condition	96	Homogeneous	100	One defect 97									Multiple defects	62
Kind of defect	88.9		100	Hole 82				Crack 88						60
Defect orientation	85.4		100	X axis 72		Y axis 75		XY plane 84		ZY plane 90		XZ plane 89		58
Defect dimension	78.1		100	P 62	HP 50	P 48	HP 59	P 93	HP 63	P 68	HP 78	P 95	HP 82	54

Abbreviations P passing through, *HP* half-passing through

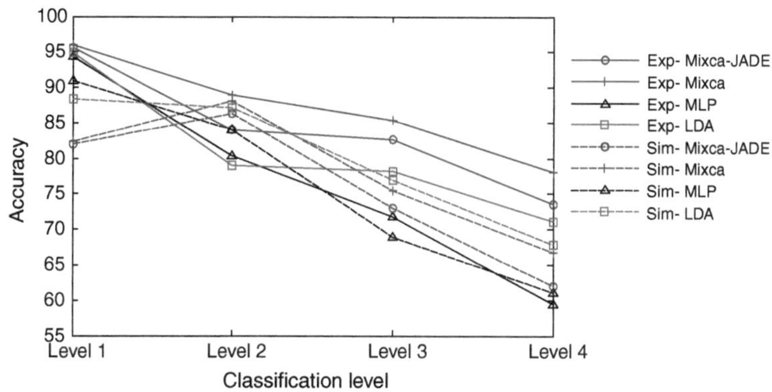

Fig. 5.11 General results of the classifiers in the different levels of classification. Abbreviations: "Exp": Experiments, "Sim": Simulations, "Mixca": non-parametric Mixca

Table 5.2 contains the confusion matrix obtained by non-parametric Mixca in the defect orientation level. The homogeneous class is perfectly classified, and the three classes corresponding to cracks are well classified. However, hole classes are frequently confused with crack classes since the defect geometry does not produce discernible wave propagation patterns. In addition, the multiple-defect class is sometimes confused with cracks. This is due to the fact that particular patterns of one of the defects inside some multiple-defect specimens are more dominant in the spectra, causing multiple-defect spectra to be similar to crack spectra.

5.4 Conclusions

ICAMM is an extension of ICA that considers a mixture of ICAs. Essentially, ICAMM is a method for versatile modelling of multivariate data densities that takes advantage of the statistical independence achieved in every ICA component. Thus, the very complex problem of modelling statistically-dependent multidimensional

Table 5.2 Confusion matrix for experiments at defect orientation level

	Homogeneous	Hole X axis	Hole Y axis	Crack XY plane	Crack ZY plane	Crack XZ plane	Multiple-defects
Homogeneous	1	0	0	0	0	0	0
Hole X axis	0	0.72	0	0.04x	0.12	0.1	0.02
Hole Y axis	0	0	0.75	0.07	0.11	0.06	0.01
Crack XY plane	0	0	0.04	0.84	0.01	0.05	0.06
Crack ZY plane	0	0.00	0.02	0	0.9	0.08	0
Crack XZ plane	0	0	0.01	0	0.1	0.89	0
Multiple-defects	0.2	0	0	0.09	0.06	0.07	0.58

data is reduced to the problem of modelling the one-dimensional data density that is associated to each possible source in every ICA member of ICAMM. Hence, in the same way that ICA is a valuable tool for blind source separation by reducing higher-order dependencies or by directly imposing statistical independence among the sources, ICAMM is also a valuable tool for classification since each ICA component can be associated to a different class of the global data model.

A new ICAMM approach has been proposed for non-destructive testing using impact-echo. The material under evaluation is modelled as a linear system that describes the wave propagation phenomenon of the impact-echo. A compressed and representative pattern of the spectra for the multichannel setup of the impact-echo has been obtained using ICAMM. This modelling allowed the spectra differences in the resonance modes to be discerned for different kinds of defective materials in several simulations and lab experiments.

We have demonstrated the feasibility of the proposed procedure for extracting patterns of different kinds of defects from impact-echo signal spectra both in simulations and in experiments. Note that there was only one piece of material for a kind of defect in a certain location in the bulk, and it was not in the training stage; therefore, the classifier had to assign the right class using the patterns of pieces of the same class in other locations. The results could be used to implement the proposed method in industrial applications of quality evaluation of materials. In these applications, the database collected within a reasonable time could have samples that are similar to the tested piece, significantly improving the classification results.

General results show that a classifier based on a mixture of independent components is a suitable technique for the impact-echo problem even in complex levels of classifications in which up to 12 classes of homogeneous, single-defect and multiple-defect materials are discerned. The underlying ICA mixture models that were learned seem to be related to the shape and location of the defects. This is a promising area of application for ICA mixture modelling.

The theoretical and experimental demonstrations provided here allow the application of the impact-echo method to be extended to the classification of different kinds of defective materials. The knowledge of the condition of the material can be enhanced, i.e., not only to detect whether a material is sound or unsound, but to obtain greater knowledge about the defects inside the material. The proposed procedure is intended to exploit to the maximum the information obtained with the cost efficiency of only one single impact. There is a range of future directions for this research, such as applying the proposed method in industrial contexts and obtaining greater insights into the sources and mixing matrices of the model in order to exploit all the information collected by the sensors. From these insights, an accurate localization of the defects and a 3D reconstruction of the internal structure of the material can be performed.

References

1. M. Sansalone, W.B. Streett, *Impact-echo: Non-destructive Evaluation of Concrete and Masonry* (Bullbrier Press, USA, 1997)
2. N.J. Carino, *The Impact-Echo Method: An Overview*, ed by P.C Chang. Structures Congress and Exposition (Washington D.C, 2001), pp. 1–18
3. A. Hyvärinen, J. Karhunen, E. Oja, *Independent Component Analysis* (John Wiley & Sons, New York, 2001)
4. A. Cichocki, S. Amari, *Adaptive Blind Signal and Image Processing: Learning algorithms and applications* (Wiley, John & Sons, New York, 2001)
5. C.F. Morabito, Independent component analysis and extraction techniques for NDT Data. Materials Evaluation **58**(1), 85–92 (2000)
6. R. Boustany, J. Antoni, Blind extraction of a cyclostationary signal using reduced-rank cyclic regression - a unifying approach. Mech. Syst. Signal Process. **22**, 520–541 (2008)
7. J. Antoni, Blind separation of vibration components: Principles and demonstrations. Mech. Syst. Signal Process. **19**, 1116–1180 (2005)
8. J.J. Gonzalez de la Rosa, R. Piotrkowski, J.E. Ruzzante, Higher order statistics and independent component analysis for spectral characterization of acoustic emission signals in steel pipes. IEEE Trans. Instrum. Meas. **56**(6), 2312–232 (2007)
9. J.J. Gonzalez de la Rosa, C.G. Puntonet, I. Lloret, An application of the independent component analysis to monitor acoustic emission signals generated by termite activity in wood. Measurement **37**, 63–76 (2005)
10. W. Li, F. Gu, A.D. Ball, A.Y.T. Leung, C.E. Phipps, A study of the noise from diesel engines using the independent component analysis. Mech. Syst. Signal Process. **15**(6), 1165–1184 (2001)
11. X. Liu, R.B. Randall, J. Antoni, Blind separation of internal combustion engine vibration signals by a deflation method. Mech. Syst. Signal Process. **22**, 1082–1091 (2008)
12. A. Salazar, L. Vergara, J. Igual, J. Gosalbez, R. Miralles, *ICA model applied to multichannel non-destructive evaluation by impact-echo*. Lecture Notes in Computer Science, vol. 3195, pp. 470–477 (2004)
13. A. Salazar, L. Vergara, J. Igual, J. Gosalbez, Blind source separation for classification and detection of flaws in impact-echo testing. Mech. Syst. Signal Process. **19**(6), 1312–1325 (2005)
14. L.K. Hansen, M. Dyrholm, A prediction matrix approach to convolutive ICA, in *Proceedings of IEEE Workshop on Neural Networks for Signal Processing XIII*, pp. 249–258 (2003)

15. M.S. Pedersen, J. Larsen, U. Kjems, L.C. Parra, *in A survey of convolutive blind source separation methods*, ed. by J. Benesty and A. Huang. Handbook of Multichannel Speech Processing Handbook, Chapter 51 (Springer, 2007), pp. 1065–1084
16. F. Meinecke, A. Ziehe, M. Kawanabe, K.R. Müller, Resampling approach to estimate the stability of one-dimensional or multidimensional independent components. IEEE Trans. Biomed. Eng. **49**(12), 1514–1525 (2002)
17. J.D. Cheeke, *Fundamentals and Applications of Ultrasonic Waves* (CRC Press LLC, USA, 2002)
18. O. Abraham, C: Leonard, P. Côte, B. Piwakowski, Time-frequency analysis of impact-echo signals: numerical modelling and experimental validation. J. ACI Mater. **97**, 6, pp. 645–657 (2000)
19. J.S. Popovics, Effects of Poisson's ratio on the impact-echo test analysis. J. Eng. Mech. **123**, 843–851 (1997)
20. A. Salazar, L. Vergara, A. Serrano, J. Igual, A general procedure for learning mixtures of independent component analyzers. Patt. Recog. **43**(1), 69–85 (2010)
21. L. Vergara, J. Gosálbez, J.V. Fuente, R. Miralles, I. Bosch, A. Salazar, A. Lopez, L. Domínguez, Ultrasonic nondestructive testing on marble block rocks. Materials Evaluation **62**(1), 73–78 (2004)
22. Y. Huang, J. Benesty, J. Chen, *Acoustic MIMO Signal Processing* (Springer,Berlin, 2006)
23. M. Joho, H. Mathis, R.H. Lambert, Overdetermined blind source separation: using more sensors than source signals in a noisy mixture, in *Proceedings of 2nd International Workshop on Independent Component Analysis and Blind Signal Separation* (Helsinki, Finland, 2000), pp. 81–86
24. C. Bailer-Jones, M. Irwin, T. Hippel, Automated classification of stellar spectra - II. Two-dimensional classification with neural networks and principal components analysis. Month. Not. R. Astron. Soc. **298**, 361–377 (1998)
25. R. Xu, H. Nguyen, P. Sobol, S.L. Wang, A. Wu, K.E. Johnson, Application of principal component analysis to the FTIR spectra of disk lubricant to study Lube–carbon interactions. IEEE Trans. Magn. **40**, 3186–3189 (2004)
26. A. Salazar, L. Vergara, R. Llinares, Learning material defect patterns by separating mixtures of independent component analyzers from NDT Sonic Signals. Mech. Syst. Signal Process. **24**(6), 1870–1886 (2010)
27. J.F. Cardoso, A. Souloumiac, Blind beamforming for non gaussian signals, in IEE Proceedings-F, vol. 140, no. 6, pp. 362–370 (1993)
28. B.W. Silverman, *Density Estimation for Statistics and Data Analysis* (Chapman and Hall, 1985)
29. A. Chen, P.J. Bickel, Efficient independent component analysis. Ann. Stat. **34**(6), 2825–2855 (2006)
30. M. Sansalone, N.J. Carino, N.N. Hsu, Transient stress waves interaction with planar flaws. Build. Res. Inf. **16**(1), 18–24 (1988)

Chapter 6
Cultural Heritage Applications: Archaeological Ceramics and Building Restoration

This chapter presents two applications: classification of archaeological ceramics and diagnosis of historic building restoration. In the first application, we consider the ICAMM-based algorithm proposed in Chap. 3 (Mixca) to model the joint-probability density of the features. This classifier is applied to a challenging novel application: classification of archaeological ceramics. ICAMM by Mixca gathers relevant characteristics that have general interest in the area of material classification. On one hand, it can deal with arbitrary forms of the underlying probability densities in the feature vector space as non-parametric methods can do. On the other hand, mutual dependences among the features are modelled in a parametric form so that ICAMM based on Mixca can achieve good performance even with a training set of relatively small size, which is characteristic of parametric methods. Moreover, in the training stage, Mixca can easily incorporate probabilistic semi-supervision (PSS): labelling by an expert of a portion of the whole available training set of samples. These properties of Mixca are well-suited for the particular problem considered: classification of ceramic pieces coming from four different periods, namely, the Bronze Age, Iberian, Roman, and the middle Ages. A set of features is obtained from the processing of the ultrasonic signal that is recorded in through-transmission mode using an ad hoc device. A physical explanation of the results is obtained with comparison with classical methods used in archaeology. The results obtained are indicative of the promising potential of ICAMM in that particular application and in the general area of material classification [1].

The second application attempts to solve two problems in NDT of historical building restoration using ICA: diagnosis of the material consolidation status and determination of the thickness of the material. In those applications, the injected ultrasonic pulse is buried in backscattering grain noise plus sinusoidal phenomena; these phenomena are analyzed by ICA. The mixture matrix is used to extract useful information concerning resonance phenomena of multiple reflections of the ultrasonic pulse at non-consolidated zones and to improve the signals by detecting interferences in ultrasonic signals. The results are shown by real experiments on a

A. Salazar, *On Statistical Pattern Recognition in Independent Component Analysis Mixture Modelling*, Springer Theses, DOI: 10.1007/978-3-642-30752-2_6,

wall of a restored dome of a Basilica. ICA is used as pre-processor to obtain enhanced power signal B-Scans of the wall.

6.1 Chronological Classification of Archaeological Ceramics

6.1.1 Introduction

Determining the historical period of archaeological ceramic shards is important for many archaeological applications, particularly to reconstruct human activities of the past. In fact, the standardization of an efficient and non-destructive testing(NDT) method for ceramic characterization could become an important contribution for archaeologists. Chemical, thermo-luminescence, and other analyses have shown to measure the age of ceramics accurately, but they are expensive, time-consuming and involve some destruction of the analyzed pieces [2]. Relative dating by comparison with ceramic collections is non-destructive but very inaccurate [2].

Ultrasound has been used in archaeological applications such as ocean exploration to detect wrecks, imaging of archaeological sites, and cleaning archaeological objects [3–5]. In this application, we consider a method to sort archaeological ceramic shards based on ultrasonic NDT. This method aims to be economic, fast, precise, and innocuous for the ceramic pieces. It consists of three steps: measuring by the through-transmission technique, extracting features from the measured ultrasonic signals, and classifying the feature set in classes corresponding to historic or protohistoric periods.

The estimation of the chronological period of an archaeological fragment is not a straightforward work, especially if we consider that the fragment might be moved from its context of origin due to migrations, wars, or trade exchange, etc. In addition, some external features used for classification of archaeological objects, such as particular shapes and decorations, might be not evident in the fragments, and thus these aspects would not provide information for a correct classification of the fragments.

Through-transmission was selected because the ceramic produces large attenuation to the propagation of ultrasound, so the pulse-echo technique cannot be implemented at the required operating frequency. Time, frequency, and statistical features (to be described later) were extracted using standard signal processing techniques. The characteristics of the classification problem offer a good case study for testing advanced classifiers, like those based on modelling the underlying statistical densities of the feature space as mixtures of independent component analyzers(ICA).

In consequence, we dedicate Sect. 6.1.2 to presenting the ultrasound through-transmission model from a linear system perspective and to defining the selected features. Then, in Sect. 6.1.3 we present the rationale for these classifiers and describe them based on mixtures of ICA. Section 6.1.4 presents the experiments and the results obtained in the sorting of ceramic pieces from four different periods: Bronze Age, Iberian, Roman, and Middle Ages. Section 6.1.5 presents the conclusions and future line of work.

We reported some preliminary results related to this archaeological application which was presented in conference [6]. The following significant new contributions are presented in this application: rationale and selection of new ultrasonic features; use of a classifier that is based on probabilistic semi-supervision of ICA mixture models that are suitable for handling expert uncertainty; implementation of an ad hoc device designed to avoid the uncontrolled conditions of a totally manual measurement procedure; and demonstration of physical interpretation of the results obtained by the proposed method in comparison with standard methods used in archaeology. Therefore, this work provides the foundations to implement a practical method to complement or even replace some of the destructive and time-consuming techniques that are currently employed in archaeology.

6.1.2 Through-Transmission Model and Feature Definition

A simplified model of ultrasonic through-transmission analysis is to consider that the recorded signal is the convolution of the material reflectivity with a linear time varying (LTV) system (see Fig. 6.1). The variant impulse response of the LTV is the injected ultrasonic pulse travelling through the material, which bears the effects of attenuation and dispersion that affect both its amplitude and frequency content. Actually, some non-linearity may be incorporated into this simple model in some specific cases; however, in general, the linear assumption is adequate for a large number of situations, or is at least enough to be able to obtain practical solutions yielding reasonable performance. Thus, the received signal $x(t)$looks similar to the one shown in Fig. 6.1.

If we consider that $x(t)$is a realization of a nonstationary stochastic process $\{\tilde{x}(t)\}$ having instantaneous power spectral density $P_x(f, t)$, different "ultrasonic signatures" $us(t)$ can be computed like those included in Eqs. (6.1–6.4).

$$\text{Centroid frequency (fc): } us(t) = f_c(t) = \frac{\int_{f_1}^{f_2} f \cdot P_x(f,t)df}{\int_{f_1}^{f_2} P_x(f,t)df} \tag{6.1}$$

$$\text{Maximum frequency(fmax): } us(t) = f_{\max}(t) = \underbrace{\max}_{f} P_x(f,t) \tag{6.2}$$

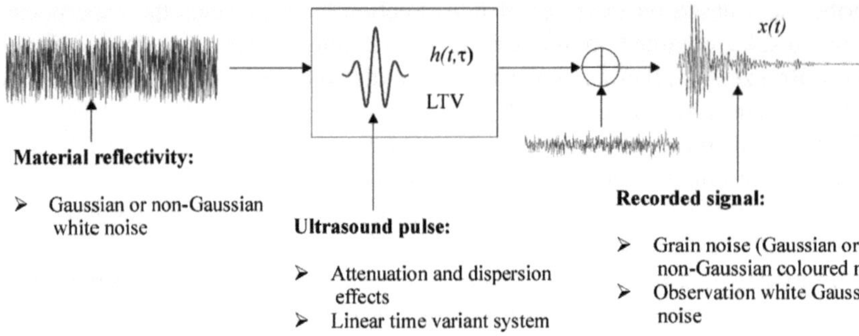

Fig. 6.1 The through-transmission linear time variant model

$$\text{Bandwidth(BW)}: us(t) = BW(t) = \frac{\int_{f_1}^{f_2} (f - f_c(t))^2 \cdot P_x(f,t)df}{\int_{f_1}^{f_2} P_x(f,t)df} \tag{6.3}$$

$$\text{Maximum frequency amplitude (Afmax)}: us(t) = \max P_x(f,t) \tag{6.4}$$

These signatures are measures of the spectral content variations that are affected by the ultrasonic pulse travelling inside the material. They can be estimated by means of well-known smoothing techniques of time–frequency spectral analysis [7].

From $us(t)$, we can obtain features in different forms. For example, the time average value $\frac{1}{t_1 - t_0}\int_{t_0}^{t_1} us(t)dt$ or the instantaneous value at one particular time instant$us(t_0)$ can be elements of the feature vector in the observation space. Other time-domain features, such as the parameters A and β corresponding to an exponential model of the signal attenuation $\hat{x}(t) = Ae^{-\beta t}$ or the total signal power received $P = \int_0^T |x(t)|^2\, dt/.T$, are also possible to complement the frequency-domain features.

More features can be defined considering special conditions of the through-transmission model. For example, higher-order statistics can be used to detect the possible degree of non-gaussianity of the reflectivity by measuring higher-order moments of the received signal like $HOM = E[x(nT_s) \cdot x((n-1)T_s) \cdot x((n-2)T_s)]$ 0, where $E[\cdot]$ means statistical expectation and $1/T_s$ is the sampling frequency. Departures from the linear model of Fig. 6.1 can be tested in different forms, for example, using the so-called time-reversibility [8], which is defined by $TR = E\left[\left(\frac{dx(t)}{dt}\right)^3\right]$.

6.1.3 Motivation for an ICAMM Application

Let us consider a probabilistic classification context where some selected features are organized as elements of vectors belonging to an observation space to be divided into K classes$\{C_k\}$ $k = 1\ldots K$. Given an observed feature vector $\mathbf{X}$, we want to determine the most probable class. More formally, we want to determine the class $C_k$that maximizes the conditional probability$p(C_k/\mathbf{x})$. Since classes are not directly observed, Bayes theorem is used to express $p(C_k/\mathbf{x})$ in terms of the class-conditioned observation probability density $p(\mathbf{x}/C_k)$ in the form $p(C_k/\mathbf{x}) = p(\mathbf{x}/C_k)p(C_k)/p(\mathbf{x})$. Note that $p(\mathbf{x})$ is a scaling factor that is irrelevant to the maximization of $p(C_k/\mathbf{x})$, and that a *priori* probability $p(C_k)$ is assumed to be known (or equal to $1/K$ for all classes). Hence, the key problem focuses on estimation of $p(\mathbf{x}/C_k)$.

A non-parametric classifier tries to estimate $p(\mathbf{x}/C_k)$ from a training set of observation vectors, but this becomes progressively intractable as the dimension of the observation space (number of features) increases, because the required size of the training set becomes prohibitive. On the other hand, a parametric classifier assumes a given form for $p(C_k/\mathbf{x})$ and, thus, tries to estimate the required parameters from the training observation set [9]. Most of the classifiers from parametric approaches consider Gaussian densities to simplify the problem in the absence of other information that could lead to better choices. Moreover, both parametric and non-parametric classifiers are very much complicated in semi-supervised scenarios, i.e., when part of the observed vectors belonging to the training set have unknown classes [10].

Therefore, procedures that would be of interest in the general area of classification should combine the following characteristics: the versatility of the non-parametric approach (from the point of view of the assumed form of $p(\mathbf{x}/C_k)$); the simplicity of the parametric methods (in the sense that most effort will concentrate on the estimation of a finite set of parameters); and operate in semi-supervised scenarios. This is especially remarkable in the area of non-destructive classification of materials. On one hand, the prediction of the joint-density of some selected features is almost impossible (Gaussianity is an assumption that is too restrictive in many cases). On the other hand, there are some applications where the available set of specimens used to obtain the training set can hardly be classified. This happens, for example, when the specimen cannot be destroyed to find the true inner state or when the definition of the K classes is not clearly known a *priori*.

The classification problem considered in this application has the conditions necessary for verifying the usefulness of a versatile classifier that is capable of working with semi-supervised training. Ceramic composition is assumed to be different in different historic and proto-historic periods, so there should be opportunities to classify the pieces from features derived from ultrasonic analysis. Nevertheless, exact modelling of the propagation of ultrasound in ceramic and statistical characterization of the features is a complex matter. Hence, it is advisable not to assume particular parametric distributions (like normal density) in the design of the classifier. On the other hand, very often, the archaeologist does not know the period of all the available specimens that could be used to form the training set of observation: semi-supervised training is a requirement in this application. Even more interesting is that the expert archaeologist can assign some probabilities of classes (ranging from 0 to 1) to part of the pieces of the training set, scenario that we will call PSS. Most of the semi-supervised classifiers are not capable of dealing with PSS, only if the assigned probabilities to the labelled feature vectors are 0 or 1. Therefore, we used the Mixca procedure explained in Chap. 3 which meets the required conditions for this classification problem.

6.1.4 Experiments and Results

Two identical transducers (one in emitter mode and the other one in receiver mode) with a nominal operating frequency of 1.05 MHz were used to obtain the through-transmission signals. This operating frequency was selected after performing different tests, as the most appropriate to achieve small ultrasound attenuation with resolution enough to separate different kinds of ceramics. Sampling frequency was 100 MHz and the observation time was 0.1 ms (10,000 samples) for every acquisition. To reduce observation noise 16 acquisitions were averaged. The size of the transducers was also important since the ceramic pieces were small (a few centimetres in height and length, and less than one centimetre in width, see Fig. 6.2).

The ceramic pieces were measured using a device where the ceramic piece is placed between two cases that adjust to the curved surfaces of the piece (see Fig. 6.3). this device was implemented to perform controlled and repeatable measurements, thereby improving the manual recording. A rubber adaptor was used as coupling medium to match the acoustical impedance of the transducer to the piece. The adaptor has a good coupling to the surface of the material and be innocuous to the piece. The emitter is located in a case on the lower side of the piece and the receiver is located in case on the upper side of the piece. Note that the transducers are embedded into a case that has a pressure control that allows the force that is applied to the material to be the same for each measurement. Since the propagation velocity is an important feature for classification, the device has a mechanism that allows that piece thickness to be measured and transmitted to the signal processing system simultaneously with the ultrasound measurement.

The distribution of the pieces was: 47 Bronze Age, 155 Iberian, 138 Roman, and 140 Middle Ages. Thus, a total of 480 pieces were used in the experiments from deposits at the Valencian Community in Spain. The features were selected from the features defined in Sect. 6.1.2. A total of 11 features were considered. The first four were the time averages over the whole acquisition interval of the 4 ultrasonic signatures defined in Eqs. (6.1–6.4). The squared magnitude of the Short Term Fourier Transform was used to estimate $P_k(f,t)$.

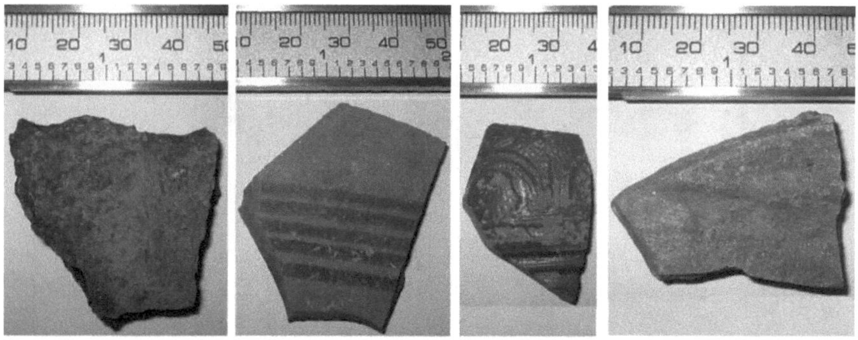

Fig. 6.2 Images of typical ceramic pieces

Feature number 5 was $f_c(t_0)$, the instantaneous value of the centroid frequency at a specific time instant. The parameters A, β, P, HOM and TR that are defined in Sect. 6.1.2, were also included in the feature vector. Finally, the velocity of propagation v of the ultrasound, which was measured by dividing the piece thickness by the pulse arrival delay, was also considered, since it is a standard variable in the ultrasonic characterization of materials. Figure 6.4 shows examples of the time record, spectrum and histogram for each period. It also shows the eleven features obtained for each example. Note the significant differences (in general) among the feature values corresponding to different periods, which provide the opportunity for good classification performance.

First, the signal features were pre-processed with PCA 0 to reduce the dimension of the problem as much as possible and to detect redundancies among the selected features. This resulted in only six significant features (components), which were linear combinations of the original ones. These 6 components explained a total of 90 % of the data variance.

Wc had a total of 480*0.75 = 360 original samples for training. By adding spherical Gaussian noise to the original samples, three replicates were estimated to obtain a total of 1,440 samples for training. We performed 100 runs varying the sets of 360 samples used for training and 120 used for testing. The percentage of success in determining the correct class was then evaluated for a total of 120 × 100 testing samples.

Different alternative ICAMM-based classifiers were implemented together with other typical classifiers. We considered four embedded ICA algorithms: non-parametric ICA (Mixca) [11]; JADE (Mixca-JADE) [12]; TDSEP (Mixca-TDSEP) [13]; and FastIca (Mixca-FastIca) [14, 15]. Several PSS ratios were also tested (PSS ratio is defined as the proportion between probabilistically labelled and unlabelled data in the training stage). Linear Discriminant Analysis (LDA) classifier 0 was also verified as it is representative of a supervised classifier optimum under Gaussianity assumptions. Some other classifiers based on neural networks schemes were also implemented: radial basis function (RBF), learning vector

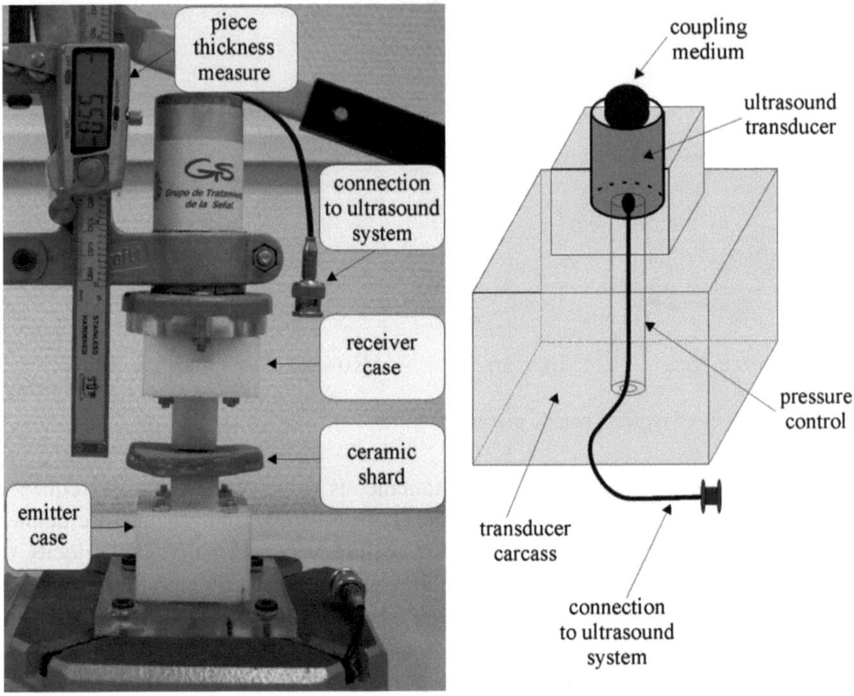

Fig. 6.3 Measurement device employed in ultrasonic signal acquisition. A detail of the ultrasound transducer case is included

quantization (LVQ), and multilayer perceptron (MLP) [16]. As well the k-nearest neighbour (kNN) was tested 0. Table 6.1 shows the overall percentage of classification accuracy achieved by the different Mixca variants.

Table 6.2 shows the overall percentage of classification accuracy achieved by the other different methods implemented. Note that different values of the fitting variables required in each method (e.g., the value k in kNN) were tested and the results shown are the best ones obtained.

The best performance in classification was obtained using Mixca at PSS ratio of 1 (total probabilistic supervision), achieving a classification accuracy of 83 %, which is much better than the rest of supervised methods (LDA, RBF, LVQ, MLP, kNN). As the PSS ratio is reduced, the performance of Mixca gets worse. However, for PSS ratio 0.8, Mixca is still the best one with classification accuracy of 79 %. For PSS ratio 0.6, only LDA gives a slightly better result. This confirms the convenience of not assuming any parametric model of the underlying probability density as is assumed in LDA and in the parametric Mixca variants. Besides, other supervised non-parametric methods (RBF, LVQ, MLP, kNN) cannot compete with Mixca since it is a hybrid method with an implicit parametric model (ICA), which allows of a training set of relatively small size.

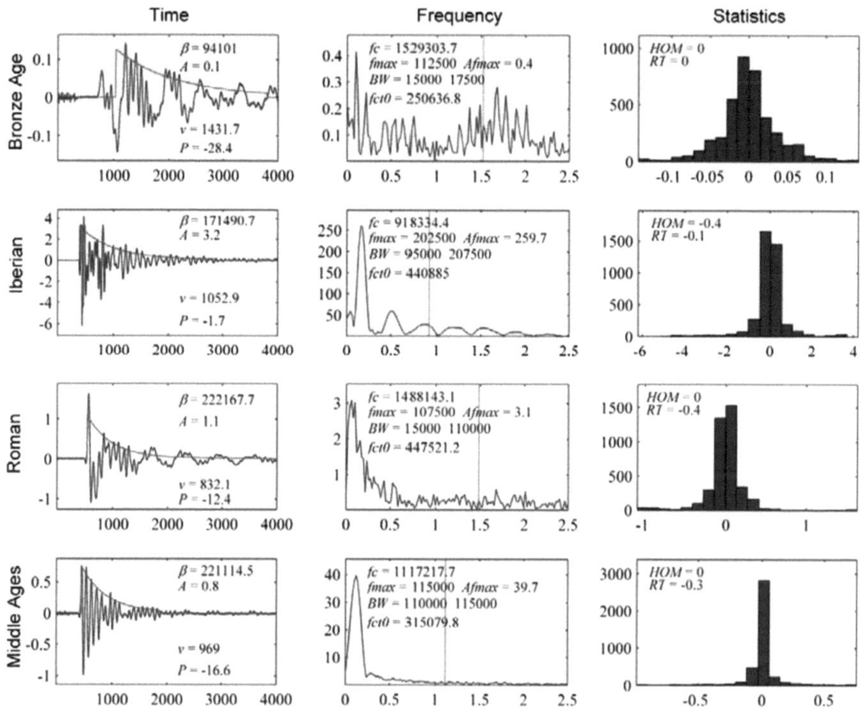

Fig. 6.4 Some examples of time signals, spectra, histograms and corresponding features extracted from ultrasonic signals for archaeological ceramic pieces from different periods. Units are: Time axis: sample number; Frequency axis : MHz; Statistics axis: bins of signal values; P [dB]; v [m/s]; fc, fmax, BW, fct0 [Hz]

Table 6.1 Classification accuracy (percentage) obtained with the different variants of Mixca

PSS ratio	Mixca	Mixca-JADE	Mixca-TDSEP	Mixca-FastIca
1	0.83	0.81	0.79	0.81
0.8	0.79	0.75	0.69	0.71
0.6	0.72	0.67	0.66	0.60
0.4	0.65	0.64	0.55	0.59

Table 6.2 Classification accuracy (percentage) obtained with the other methods

LDA	RBF	LVQ	MLP	kNN
0.73	0.64	0.59	0.67	0.64

To gain more insight into the classifier performance, we include Table 6.3, which contains the confusion matrix obtained by Mixca for 1 PSS ratio. The Roman and Iberian categories are not very difficult to classify, but they are often confused with each other. The pieces from the Middle Ages are confused with

Table 6.3 Confusion matrix (percentages) by Mixca with 1 PSS ratio

	Bronze age	Iberian	Roman	Middle ages
Bronze age	0.79	0	0.07	0.14
Iberian	0	0.89	0.09	0.02
Roman	0.05	0.19	0.69	0.07
Middle Ages	0.02	0	0.05	0.93

Bronze Age pieces 14 % of the time, and Roman pieces cause misclassification of some pieces from the Bronze Age and the Middle Ages.

6.1.5 Discussion

In order to draw a physical interpretation of the results obtained by ultrasounds, a diversity of morphological and physiochemical characterization analyses were carried out using conventional instrumental techniques. A stratified random sampling analysis was made using data from the physical analysis of the pieces: open porosity and apparent density [17, 18]. Thus, a sample of the ceramic pieces for the different periods was obtained. The raw material composition of the selected pieces was analyzed using optical microscope and scanning electron microscope (SEM) [19, 20]; and also the processing methods of the pieces were studied. From those analyses, the differences of the ceramic physical properties for the different periods and the ultrasound propagation are discussed.

6.1.5.1 Open Porosity and Apparent Density

A sample of the pieces was selected for morphological and physiochemical characterization based on open porosity and apparent density analyses of the pieces. For stratified random sampling, the values of these physical properties for the different periods were considered as random variables that follow Gaussian distribution. First, an estimation of the variable variance for the different periods (statistical strata) was made. This estimation was obtained from 45 representative pieces that were physically tested for open porosity and apparent density. The results of this prior study are shown in Table 6.4.

The objective of the sampling was to provide estimators with small variances at the lowest cost possible (considering that morphological and physiochemical characterization are costly).To estimate the fraction of the total sample size n corresponding to the stratum i, we applied the so-called Neyman allocation [21], $\frac{n_i}{n} = \frac{N_i \sigma_i}{\sum_{i=1}^{L} N_i \sigma_i}$, where L is the number of strata (4 periods for this application), N_i is

Table 6.4 Porosity and density statistics of the prior study

	Open porosity (%)		Apparent density (gr/cm3)	
Period i	Mean $\hat{\mu}$	Standard deviation $\hat{\sigma}_i$	Mean $\hat{\mu}$	Standard deviation $\hat{\sigma}_i$
1. Bronze Age	28.20	3.7794	1.80	0.0676
2. Iberian	22.70	3.3320	1.85	0.0663
3. Roman	31.06	8.3532	1.79	0.1607
4. Middle Ages	22.69	5.3441	1.84	0.0949

the sample number in the stratum i(3, 15, 15, 12 for Bronze Age, Iberian, Roman, and Middle Ages pieces, respectively), and σ_i is the standard deviation for the stratum i(estimates of Table 6.4 were applied). The results for the strata $i = 1...4$ were: $\frac{n_1}{n} = 6.85\,\%$, $\frac{n_2}{n} = 19.90\,\%$, $\frac{n_3}{n} = 44.42\,\%$, and $\frac{n_4}{n} = 28.83\,\%$ for open porosity; and $\frac{n_1}{n} = 6.5\,\%$, $\frac{n_2}{n} = 21.01\,\%$, $\frac{n_3}{n} = 45.34\,\%$, and $\frac{n_4}{n} = 27.16\,\%$ for apparent density, respectively.

We specified that the estimate of the sample mean should lie between B units of the population mean, with probability equal to 0.95. This is equivalent to impose that the mean estimate should lie in the interval $\mu \pm 2 \cdot \sigma$, i.e., $B = 2 \cdot \sigma$. From the analysis of the variable means of Table 6.4, we chose $B = 1.1\%$ and $B = 0.02$ gr/cm^3 as the bounds on the error of estimation of the population mean for open porosity and apparent density, respectively. These bounds allowed the stratum mean of the sampling to be separated adequately.

The total number of samples was estimated using 0, $n = \frac{\left(\sum_{i=1}^{L} N_i \sigma_i\right)^2}{N^2 \cdot D + \sum_{i=1}^{L} N_i \sigma_i^2}$, where $D = B^2/4$. Thus, we obtained the total number of samples $n = 79$ and $n = 83$ for open porosity and apparent density, respectively. These were the number of pieces that the morphological and physiochemical characterization analyses were applied to. Using the estimated fractions $\frac{n_i}{n}$ for the strata and the total number of samples n, we obtained the sampling population for each stratum. The final results of the stratified random sampling for an error margin of 0.05 are in Table 6.5. The estimate of the population mean for open porosity and apparent density for each stratum are shown with an approximate 2 standard deviation bound on the error of estimation.

Table 6.5 shows that the samples of the different strata (chronological periods) can be clearly separated by open porosity, since the bounds of the distributions define the most part of the densities to be disjoint. The separation of the samples by apparent density is more difficult because there is a degree of overlapping between densities of Roman and Bronze Age pieces, and a higher overlapping between densities of Iberian and Middle Ages pieces. However the joint densities of these two collections of pieces are well-separated between them. In conclusion, physical

Table 6.5 Statistics of the stratified random sampling for open porosity and apparent density

		N_i	n_i	μ_i	$\mu_i \mp 2 \cdot \sqrt{\left(\frac{N_i - n_i}{N_i}\right)\left(\frac{\sigma_i^2}{n_i}\right)}$	
Open porosity	1. Bronze Age	47	5	29,30	27,70	30,90
	2. Iberian	155	16	22,50	21,71	23,29
	3. Roman	138	35	32,00	30,78	33,22
	4. Middle Ages	140	23	23,80	22,78	24,82
Apparent density	1. Bronze Age	47	5	1,85	1,82	1,88
	2. Iberian	155	17	1,77	1,75	1,79
	3. Roman	138	38	1,87	1,85	1,89
	4. Middle Ages	140	23	1,78	1,76	1,80

properties of the ceramics shows that it is possible a separation of the pieces in the different chronological periods of this study. Different porosities and densities of the pieces are determined by the material composition and processing methods employed in the ceramic manufacturing. These issues are studied in the next section.

6.1.5.2 Ceramic Composition and Processing

The selected pieces were observed, photographed, and then analyzed using an optical microscope and a scanning electron microscope (SEM). Some of the test tubes prepared for SEM are shown in Fig. 6.5.

The data provided by optical microscope and SEM show that there are clear differences at a morphological level between the different groups of processed fragments. Therefore, the ceramic pieces corresponding to the Bronze Age exhibited a dark brown tone and the presence of a lot of dark-toned ferrous-composition spots that are associated with magnetite as well as reddish ferrous iron oxide nuclei. The Iberian ceramic pieces had varying shades between orange and black. The quartz temper was big or very big grains and abundant ferrous iron oxide nuclei as well as more isolated dark magnetite spots were found. This was an iron-rich ceramic (up to 7.45 % of FeO_3) with a high content of calcium (up to 6.30 % of CaO). The fragments of Roman ceramic had variable characteristics depending on the typology (sigillata, common, and amphora). In any of these, the pieces were made of an orange-toned paste with small-size porosity and small quantity of temper that increased from the amphora to the sigillata typology. Roman ceramic showed content of Fe_2O_3 of 5.71, 6.36 and 9.24 %, and content of CaO of 0.67, 2.92 and 1.29 % for sigillata, common and amphora, respectively. Finally, the ceramic from the Middle Ages had a bright orange to brown colour that indicates they are made of ferrous paste. This ceramic contains abundant small to very small nuclei of red ferrous iron oxide as well as dark-toned magnetic spots and quartz temper of big or very big grains. Also, limestone aggregates of white tone associated with high content of CaO (around 8 %) were observed.

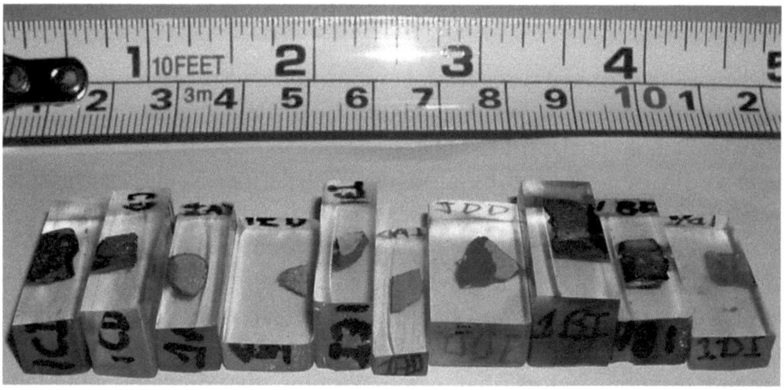

Fig. 6.5 Bits taken from the ceramic fragments included in the test probes prepared for the Scanning Electron Microscope

With regard to the methods used to manufacture the ceramics, they were different according to the evolution in time of the processing techniques. The set of ceramic fragments were from three regions (Requena, Enguera, and Liria) from the Valencia Community at the East of Spain. The pieces of the Bronze Age were from Requena (XXX-XX centuries B.C.). They were handmade using basic appliances, with an appearance very coarse, rudimentary, and of irregular texture for household. Manufacturing was local and authentic of every town; it was related to the women's domestic activities. From the dark tone of the Bronze Age ceramics, it can be inferred that they were made in reducing atmosphere, i.e., in closed oven at low temperatures. Iberian fragments corresponded to brush-decorated with geometric, zoomorphic, and human motifs or non-decorated vessels. These pieces have been dated at about V-III centuries B.C and they were from three different deposits. Paste of the Iberian ceramics was much more fine and elaborated than the Bronze Age ceramic paste. The technological innovation in the processing of the pieces was the use of lathe.

The Roman fragments of the three groups (sigillata, common, and amphora) showed technical perfection of manufacture using different techniques: lathe, handmade, and mold. They were from I-III centuries. In this period, the applications of molds for potters allowed mass production of ceramics. Sigillata ceramic features a red bright varnish that is obtained applying a clay solution to the ceramic surface and cooking at high temperatures in open oven (oxidizing atmosphere). Sigillata pieces were decorated with reliefs of different motifs and were luxury ceramic. Common and amphora types of Roman ceramic were made using lathe. They were rough appearance without decoration and for household and/or storage or transport use. The Middle Ages pieces were of two subperiods: Islamic and Christian (around VIII-X centuries). The Islamic pieces were from caliphate vessels of paste simple elaborated without decoration and special treatment. The Christian pieces were white gross paste of diverse typologies, some without decoration and some with incisions or decorations in black painted with manganese oxide.

6.1.5.3 Ceramic Physical Properties and Ultrasound Propagation

The differences in physical properties, composition and processing of the ceramic pieces, presented above, suggest the possibility of devising non-destructive techniques for archaeological ceramic classification. In Sect. 6.1.5.1 was shown that the pieces could be separated by chronological periods using measures of their open porosity and apparent density. Besides, it is well-known that porosity and density of a material have a definite influence on the propagation of the ultrasound [22, 23]. Thus, it is clear that should be there correlation between the results obtained by the proposed method based on ultrasounds (Sect. 6.1.4) and the differences in physical properties of the pieces for the different chronological periods.

There are several factors that can determine the porosity and density of ceramics, such as the raw material composition and the processing method employed to manufacture the pieces. However, in the case of archaeological ceramics, the original ceramic physical properties after manufacturing, can be altered by other factors such as the ceramic use (i.e., over-heating for cooking, etc.) and in general with the pass of the time (i.e., fractures, loss of cover layers, etc.). Thus, an exhaustive analysis of physical properties and how these properties were derived for archaeological ceramics becomes a very complex problem that needed an important amount of information that is outside the scope of this work. Note that the objective of this work is to provide a new NDT procedure to classify archaeological ceramics from the basis of training with a set of pieces of known class made with the intervention of an expert. A correct training will determine the achievement of the procedure to classify ceramics of unknown class.

The analysis of the results obtained by ultrasounds provided here consider correct (or at least probabilistic) labelling made by the expert and are based on available data of the composition, processing and physical features of the ceramics shown in Sects. 6.1.5.1 and 6.1.5.2. Let us explain the misclassifications in the confusion matrix of the ultrasound-based classification of Table 6.3. Misclassification is obtained from similar responses of pieces from different periods to the ultrasounds. Table 6.3 shows that Roman ceramics is the most misclassified group. Confusion between Roman and Iberian pieces (19 and 9 %) can be explained from ceramic composition and processing. The amphora and common Roman pieces were made from iron-rich paste and using lathe as well as the Iberian pieces. Thus, the mechanical and physical properties for these two groups were similar.

The confusion between Roman and Bronze Age pieces (5 and 7 %) can be explained due to changes in the structure of some of the Roman pieces of the sigillata subgroup that had lost the cover varnish. The high value of porosity shown by the fragments of sigillata is associated with pores of very small size and very connected, which allows big water absorption once the varnish is removed. Thus, these two groups of pieces show similar physical properties due accidents cause with the pass of the time. Regarding to the confusion between Bronze Age and Middle Ages pieces (14 and 2 %), this also can be explained from composition and processing. The Islamic subgroup of Middle Ages pieces were from the "paleoandalusí" period (early centuries of the Islamic period in Spain). During, this period, the productive strategy of household chose intentionally to simplify the production process. Simple ways of ceramic manufacture and cooking were employed to obtain kitchen's recipients with thermal shock resistance. Thus, ceramics were manually made from little-decanted clays and cooked at low temperatures. The results were coarse pieces from the Middle Ages with physical properties comparable to the Bronze Age pieces [24].

Let us analyze the ultrasound-based results from the point of view of the porosity and density. We observed that the porosity and density of the Bronze Age pieces are relatively close to porosity and density of the pieces from Roman and Middle Ages. This explains why 7 and 14 % of the Bronze Age pieces were assigned to the Roman and Middle Ages periods in Table 6.3. Similarly, the pieces from the Iberian period and the Middle Ages have similar porosities and densities, so this may justify why 2 % of the Iberian pieces were assigned to the Middle Ages. The 9 % of pieces of the Iberian period that should have been assigned to the Roman period were incorrectly assigned because the Iberian ceramic is very close to one of the three kinds of Roman ceramics (sigillata, common, and amphora)—the common kind—. This also explains why the corresponding 19 % of pieces of the Roman period were incorrectly assigned to the Iberian period. No clear explanation exists for the lack of symmetry in the confusion matrix of Table 6.3; however, it must be taken into account that the training process introduced some degree of arbitrariness because of the probabilistic labelling of the expert. Thus, it seems that the expert was able to clearly identify the pieces from Iberian and Middle Ages, but had more difficulties with the Bronze Age and Roman ones. This uncertainty may have been transmitted to the classifier during the training stage.

The experiments with standard methods of ceramic characterization used in archaeology not only show that correlations between the extracted parameters from the ultrasound signals and the physical properties of the materials were found. Moreover, they also have demonstrated some advantages of the proposed ultrasound method. The equipment required for NDT by ultrasound is, in general, less costly, and the experiments are easier to perform. The pieces are not damaged in any way during testing, nor is it necessary to alter or destroy any of the material that is analyzed. Very significant differences for the time required to analyze the pieces were demonstrated: the ultrasound analysis (measuring, processing, and automatic classification) for 480 pieces took only 6 h; the SEM analysis (tube preparation and electron microscope analysis) for 80 pieces took 274 h; the porosity and density analyses (immersion and weighing of the pieces) for 80 pieces took 288 h.

There are limitations to the application of this procedure due to the fact that the training of the classifier must be performed from a specific set of data. Thus, the classifier must be adapted to a specific data model and its efficiency is restricted by the fact that the new data to be classified must follow the same data model. Nevertheless, the training of the classifier could progressively be improved by increasing the number of pieces for each known chronological period. With proper training, the classifier would be able to provide a prediction of the chronological period for pieces that do not have clear chronological markers. In addition, the semi-supervised training mode could be used to model the uncertainty that expert archaeologists may have about the chronological period to which the pieces belong.

6.2 Consolidation Diagnosis and Layer Determination in Heritage Building Restoration

6.2.1 Introduction

NDT by ultrasounds is a very useful technique applied in fields such as construction, food, and biomedicine. The technique has basically two operation modes: pulse-echo (one sensor as emitter and receiver) and transmission (one emitter and one receiver). An ultrasound pulse is injected in the inspected material and a response of the material microstructure is received [22, 23]. The measured signal can contain echoes produced from discontinuities, inhomogeneities, borders of the material, plus material grain noise (superimposition of many small echoes due to the material microstructure). All of this information can be used for quality control and characterization of materials [25, 26]. The present study used the pulse-echo technique, given that the inspected material consisted of a wall with no possible access from opposite sides. This wall was a zone at the dome of the *Basilica de la Virgen de los Desamparados* in Valencia, Spain.

This section includes two novel applications of ICA [27, 28] as pre-processor in ultrasound NDT applied to historical building restoration. The first application consisted in using the mixture matrix to distinguish information about the consolidation process of the material. This was done by injecting a product to fill cracks in the wall. The second application consisted in detecting interferences in the recorded signals to cancel them thereby improving the quality of the signals. This procedure was applied to recorded signals for estimating the thickness layer profile of the wall.

Interferences can be due to the internal clocks of the measurement equipment, interferences with other equipment, and so on. In many applications, the recording of high quality raw data is a difficult task, especially in situations where the conditions cannot be controlled during the experiment. One difficulty in obtaining the measurements at the dome was the use of a plastic for covering the transducer in order to avoid direct contact of the ultrasonic coupling gel with the artistic paintings on the walls. This kind of measurement produced attenuated signals.

B-Scans diagrams were used to visualize consolidated and non-consolidated material zones to check the quality of restoration and to detect interfaces between the different materials in the wall. B-Scan is a 2D representation of a signal set. The evolution in time windows of a parameter such as power or spectrum was calculated for each one of the signals. Then, all of the calculated information was

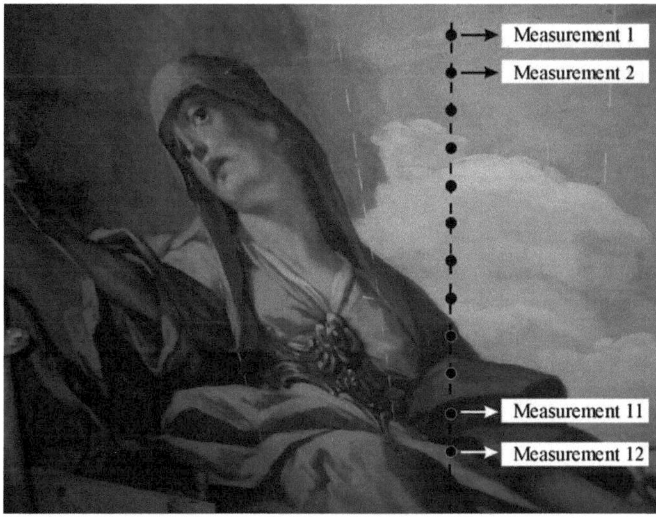

Fig. 6.6 Ultrasound inspection at the dome

put together in a representation of the measurement point versus the signal parameter evolution. Figure 6.6 shows different points of a B-Scan measured by ultrasound at the dome.

The following sections describe: the ICA model of the problem; the performed experiments, including the equipment setup; a comparison between the B-Scans that were obtained with and without ICA as a pre-processor; and an analysis of the sensitivity to detect interferences.

6.2.2 ICA Statement of the Problem

The recorded signals are modelled as the superposition of the backscattered signal plus sinusoidal phenomena. This sinusoidal contribution should be determined in order to know whether or not it is due to useful information about the material structure such as material resonances, or whether it is due to interferences from the instrumentation during measurement. The ICA statement of the problem is:

$$x_k(t) = s_k(t) + \sum_{i-1}^{N-1} \alpha_{ik} e^{j(\omega_i t + \theta_{ik})} \quad K = 1 \ldots M \tag{6.5}$$

where M is the number of measurements, $x_k(t)$ is the signal received from the material at the position k of the B-Scan, $s_k(t)$ is the backscattering signal that depends on the material microstructure, and $\alpha_{ik} e^{j(\omega_i t + \theta_{ik})}$ $i = 1 \ldots N - 1$, $k = 1 \ldots M$ are the sinusoidal sources to be analyzed.

The backscattering signal, under certain assumptions related to the wavelength of the ultrasonic wave and the scattering size, can be modelled as a stochastic process given by:

$$\{\tilde{Z}(\mathbf{x},t)\} = \sum_{n=1}^{N(\mathbf{x})} \tilde{A}_n(\mathbf{x})f(t - \tilde{\tau}_n(\mathbf{x})) \tag{6.6}$$

where $\mathbf{x}$ is the transducer location (we obtain different backscattering registers for different transducer locations). The random variable $\tilde{A}_n$ is the scattering cross-section of the nth scatter; the random variable $\tilde{\tau}_n$ is the delay of the signal back-scattered by the nth scatter; and $N(\mathbf{x})$ is the number of scatters contributing from this position. The function $f(t)$ is a complex envelope of the ultrasonic frequency pulse, that is

$$f(t) = p(t)e^{j\omega_0 t} \tag{6.7}$$

where $p(t)$ is the pulse envelope and ω_0 the transducer central frequency.

The backscattering model of Eq. (6.6) is composed of a homogeneous non-dispersive media and randomly distributed punctual scatters depicting the composite nature of the received grain noise signal instead of a rigorous description of the material microstructure [29].

In the simplest case consisting of a homogeneous material and only one harmonic of the sinusoidal components, the ICA model of Eq. (6.5) is.

$$x_k(t) = s(t) + \alpha_k e^{j(\omega_i t + \theta_k)} \quad k = 1\ldots M \tag{6.8}$$

As is well-known, standard ICA (no prior information ICA model included) requires as many mixtures as sources. In the case of Eq. (6.5), a B-Scan of 2 points would be enough. In the proposed applications, $M = 12$ and 10, therefore 12 and 10 mixtures were used to include the anomalies of the material and allow a relatively high number of interferences. Even if there are not enough points with the M points registered, the number of sensors can be virtually increased if responses to different pulses are recorded, considering that the echo is the same and the pulse repetition period is not a multiple of the sinusoid period [30, 31].

Obviously the sinusoidal components have the same frequencies throughout the B-Scan, with possibly changing amplitude and phase. From a statistical point of view, considering the interference or resonance as a sinusoid with deterministic but unknown amplitude and uniform random phase, it is clearly guaranteed that the backscattering signal and it are statistically independent.

The objectives of the experiments were to visualize non-consolidated zones and to calculate layer thickness at the wall of the dome. Ultrasound transducers have a working transmission frequency: the higher the transducer frequency, the higher the capacity to detect small details (but also the lower the capacity of material penetration). Therefore, smaller details can be detected using high frequency transducers, but they have to be closer to the material surface. The transducer used for consolidation analysis (application 1) was 1 MHz and the transducer used for

Table 6.6 Equipment setup

Ultrasound equipment setup		Acquisition equipment setup	
Ultrasound equipment	Matec PR5000	Acquisition equipment	Oscilloscope TDS3012 Tektronix
Transducers	1 MHz (application (1) 5 MHz (application (2)	Sampling frequency	10 and 250 MHz
Pulse width	0.9 μs	Sample number	10,000
Pulse amplitude	80 %	Observation time	1 ms and 40 μs
Analog filter	200 kHz–2.25 MHz (consolidation diagnosis) 1 MHz–7 MHz (layer determination)	Vertical resolution	16 bits
Excitation signal	Tone burst 1 and 5 MHz	Dynamic range	±2.5V
Operation mode	Pulse-echo	Average	64 acquisitions
Amplifier gain	65 dB	PC connection	GPIB

the thickness layer profile (application 2) was 5 MHz. This last transducer was selected because we were interested in obtaining information of the superficial layers.

The equipment setup used for NDT of the historical building is described in Table 6.6 (an outline of equipment connections is in Fig. 6.7): The Mixca algorithm described in Chap. 3, which was configured to estimate one ICA, was used (see Appendix A). The mixture matrix obtained by ICA was used to separate the information concerning the sinusoidal phenomena.

6.2.3 Diagnosis of the Material Consolidation Status

BSS by ICA was selected for this application because, contrary to the classic spectral analysis techniques [32], BSS is an unsupervised method that does not require any estimation of the noise autocorrelation matrix in data corrupted by the sinusoidal interference, considerations on the kind of noise, or model assumptions such as the filter order in model-based methods.

Figure 6.8 shows the B-Scan estimated by signal power using a conventional non-stationary analysis applying a moving window over the 12 ultrasonic recorded signals. Figure 6.8a shows two clearly differentiated zones; the first zone corresponds to the consolidated zone (low level of signal) and the second zone corresponds to the non-consolidated zone (high level of signal). The signal penetrates well into the wall at the consolidated zone and is attenuated before reflecting any kind of signal. Conversely, the signal level is increased in a non-consolidated zone due to multiple reflections of the ultrasonic pulse (see Fig. 6.8b).

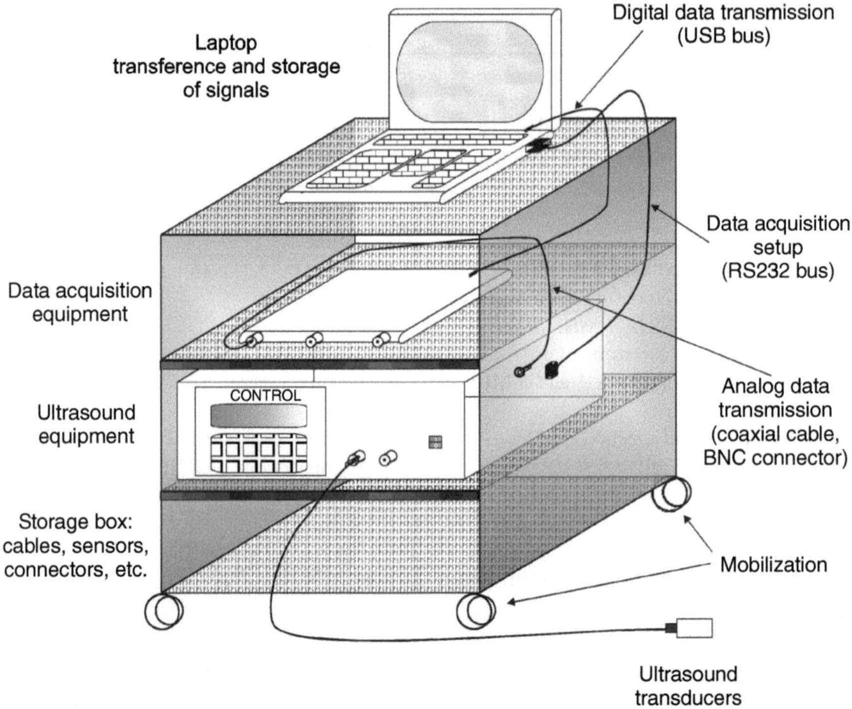

Fig. 6.7 Ultrasound equipment

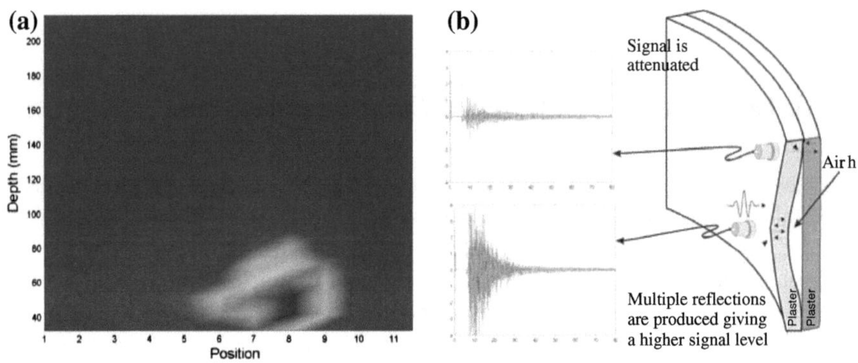

Fig. 6.8 Power signal B-Scan by non-stationary analysis **a** Power B-Scan **b** Scheme of the wall

From the spectral analysis, two frequencies (181 and 356 kHz) were found in all the recorded signals. After estimating the B-Scan of Fig. 6.8, the origin of the sinusoidal frequencies was not clear enough; they could be interferences or material resonances. Then we applied ICA to obtain more information from the mixture matrix and recovered sources. Figure 6.9 shows the recorded signals and

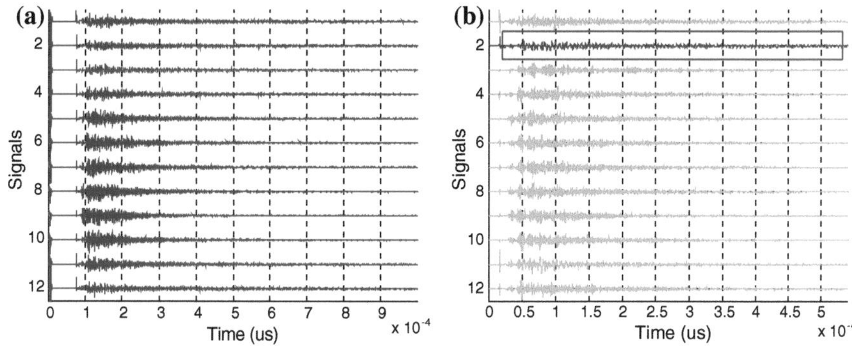

Fig. 6.9 Recorded signals and recovered sources (the supposed "interference" is highlighted) **a** Recorded signals **b** Recovered sources

the recovered sources by ICA; the sample numbers processed were from 600 to 6000.

Figures 6.10a and b show two B-Scans obtained from the mixture matrix corresponding to $\mathbf{x} = \sum_{i=1}^{12} \mathbf{a}_i, s_i, s_i = 0\ (i \neq 2)$ and $\mathbf{x} = \sum_{i=1}^{12} \mathbf{a}_i, s_i, s_i = 0\ (i = 2)$, respectively. The first B-Scan represents the sinusoidal phenomenon depicting the non-consolidated zone. Thus, this phenomenon can be associated with the shape of the material in the non-consolidated zone. The second B-Scan is the information related to the consolidated zone. The diagrams obtained from the ICA information depict the two different zones of the material more precisely than the B-Scan obtained by non-stationary analysis.

6.2.4 Thickness Material Layer Profile

Figures 6.11a and b show the recorded signals plus 1 MHz artificial interference added and the corresponding B-Scan calculated by the evolution of the centroid frequency [33]. The following information is represented in the diagram: (i) the x axis, which is the transducer position from position 0 to 10; (ii) the y axis, which is the depth axis; and (iii) the z axis, which is depicted by colours that denote the parameter level at a given position in a given depth.

The depth is obtained by $depth = velocity * time/2$ where factor 2 is due to the round trip travel of the ultrasound pulse between the material surface and the layer. The first two layers of the dome wall were composed of mortar and plaster, respectively. For the calculation of depth, an average ultrasound propagation velocity of 1600 m/s was calculated from lab probes. Due to the 1 MHz interference, the B-Scan is not clear enough to represent a profile of a layer.

Figures 6.12a and b show the results obtained by applying ICA on the ultrasonic signals. To assess the sensitivity of ICA in detecting the interference, a

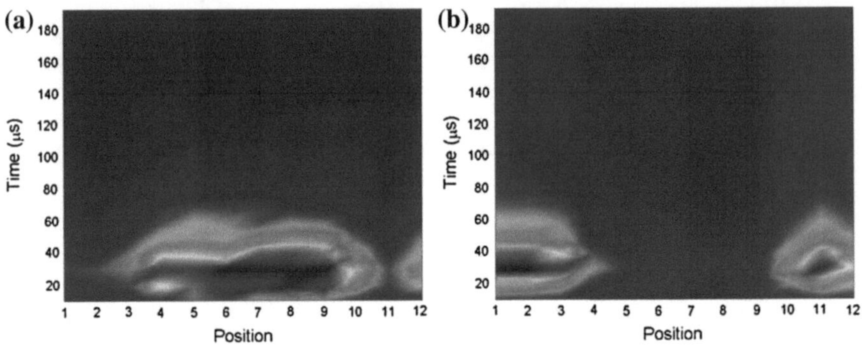

Fig. 6.10 Power B-Scan after ICA preprocessing **a** B-Scan from sinusoidal components **b** B-Scan from backscattered components

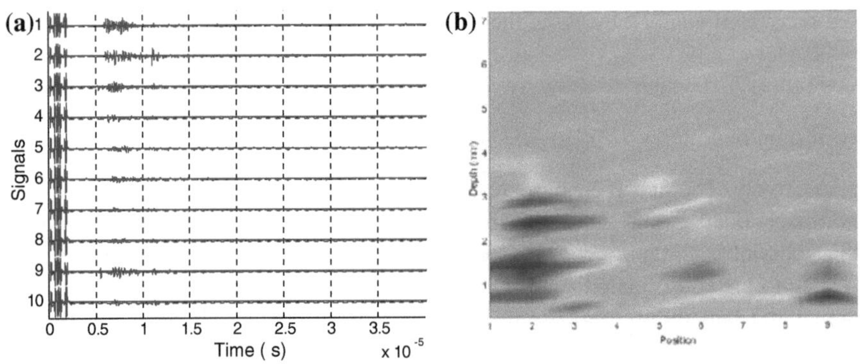

Fig. 6.11 Recorded signals and centroid frequency B-Scan by non-stationary analysis **a** Recorded signals **b** Centroid frequency B-Scan

controlled interference was added to the signals testing different frequencies and amplitudes of the interference. Figure 6.12a shows the error in the extraction of the interference versus the ratio power interference to the power signal (Pinterference/Psignal) for different interference frequencies. The higher the interference amplitude, the better the extraction of the interference; the higher the interference frequency, the worse the extraction of the interference. Figure 6.12b depicts an enhanced centroid frequency B-Scan (cancelled interference) with a layer that is clearly defined at 4 mm. corresponding to the mortar layer of the dome wall.

Other alternatives for sinusoid extraction are based on the use of the so-called notch filtering [34]. These alternatives can be designed assuming prior knowledge of the pulsations to be cancelled. Thus, BSS could be used as a prior step to notch filtering; however, it implies transient effects, possible instability problems, and some distortion of the obtained interference-free records (because of the finite notch bandwidth).

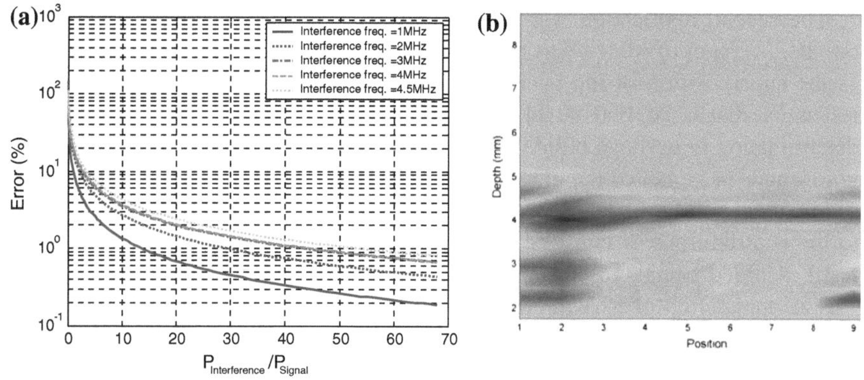

Fig. 6.12 **a** Error percentage vs. Pinterference/Psignal ratio; **b** Enhanced centroid frequency B-Scan with cancelled interferences

The results obtained from the Basilica's dome inspection were validated by testing in an architectonic scale model replica. In addition, some material samples were extracted from the replica and measured in the lab to obtain an accurate calculation of material wave propagation.

6.3 Conclusions

In this chapter, two applications of the proposed methods to NDT have been presented. In the first application, we obtained the results of applying the Mixca procedure to a challenging application in the area of non-destructive testing of materials: the classification of archaeological ceramic pieces into different historic periods. We have demonstrated the interest of using methods that are able to consider non-gaussian models of the underlying probability densities in the feature vector space. Thus, the Mixca algorithm was tested using different variants depending on the embedded ICA algorithm. Mixca has the additional merit of allowing PSS labelling, which is of practical interest in the considered application. Note that in any Mixca variant, the mutual dependence among features is modelled in a parametric form; also note that in non-parametric Mixca, the estimated marginals are non-parametric. This confirms that non-parametric Mixca shares the good general modelling capability of non-parametric classifiers and can also work with a training set of relatively small size, which is a relevant property of parametric techniques. This explains why Mixca has shown the best results and is able to produce acceptable performance for even low ratios of PSS.

The experiments show promising results in defining a standardized method that could complement or replace destructive, costly, and time-consuming techniques, which are currently being used by archaeologists in the area of ceramic characterization. Extensions of the procedures presented in this work to other emergent material applications are planned for future work.

The second application was the restoration of historical buildings where the one-ICA version of Mixca was applied. The ICA model for ultrasound evaluation as the superposition of the backscattered signal plus sinusoidal phenomena was tested by means of two novel applications: consolidation diagnosis and layer determination in heritage building restoration. The application of ICA to NDT by ultrasounds has enabled the diagnosis of the consolidation status in the restoration of historical buildings. The proposed procedure allowed the sources corresponding to the contributions of consolidated and non-consolidated zones in the backscattered recorded signals to be separated.

The application of ICA to NDT by ultrasounds made the determination of the thickness of the material profile possible and allowed interferences from the recorded signals to be cancelled. The enhanced B-Scan enabled the thickness of the first layer of mortar to be determined. ICA works well in the case of a relatively high interference level with respect to the ultrasonic signal.

Enhanced power and centroid frequency B-Scans were obtained using ICA as preprocessor of the non-stationary analysis. Future work is being addressed to apply the ICA for classification and characterization of materials.

References

1. A. Salazar, L. Vergara, ICA mixtures applied to ultrasonic non-destructive classification of archaeological ceramics. EURASIP J. Adv. Signal Process. vol. 2010, Article ID 125201, 11 pages, doi:10.1155/2010/125201 (2010)
2. R.E. Taylor, M.J. Aitken, *Chronometric Dating in Archaeology*, Advances in archaeological and museum science series, vol 2 (Springer-Verlag, New York USA, 1997)
3. R. Cribbs, F. Saleh, An ultrasonic based system used for non-destructive imaging of archaeological sites, in *Proceedings of Informatica ed Egittologia all'inizio degli anni'90*, (Roma Italy, 1996), pp. 97–108
4. A. Murray, M.F. Mecklenburg, C.M. Fortenko, R.E. Green, Detection of delaminations in art objects using air-coupled ultrasound, in *Proceedings of Materials Issues in Art and Archaeology III*, (San Francisco USA, 1992), pp. 371–378
5. W.I. Sellers, Ultrasonic cave mapping. J. Archaeological Science **25**(9), 867–873 (1998)
6. A. Salazar, R. Miralles, A. Parra, L. Vergara, J. Gosalbez, Ultrasonic signal processing for archaeological ceramic restoration, in *Proceedings of 31st IEEE International Conference on Acoustics, Speech, and Signal Processing, ICASSP*, (Toulouse, France, 2006), pp. III-1160–1163
7. L. Cohen, P. Loughlin, *Recent Developments in Time-Frequency Analysis* (Springer-Verlag, New York USA, 1998)
8. R. Miralles, L. Vergara, A. Salazar, J. Igual, Blind detection of nonlinearities in ultrasonic grain noise. IEEE Trans. Ultrasonics, Ferroelectrics, and Frequency Control **55**(3), 637–647 (2008)
9. A.K. Jain, Statistical pattern recognition: A review. IEEE Transactions on Pattern Analysis and Machine Intelligence **22**(1), 4–37 (2000)
10. O. Chapelle, B. Schölkopf, A. Zien, Semi-supervised Learning, MIT Press, 2006.
11. A. Salazar, L. Vergara, A. Serrano, J. Igual, A general procedure for learning mixtures of independent component analyzers. Pattern Recognition **43**(1), 69–85 (2010)

12. J.F. Cardoso, A. Souloumiac, Blind beamforming for non gaussian signals. IEE Proceedings-F **140**(6), 362–370 (1993)
13. A. Ziehe, K.R. Müller, TDSEP- an efficient algorithm for blind separation using time structure, in *Proceedings of the 8th International Conference on Artificial Neural Networks, ICANN'98, Perspectives in Neural Computing*, pp. 675-680 (1998)
14. A. Hyvärinen, E. Oja, A fast fixed-point algorithm for independent component analysis. Neural Computation **9**(7), 1483–1492 (1998)
15. A. Hyvarinen, Fast and robust fixed-point algorithms for independent component analysis. IEEE Trans.Neural Netw. **10**(3), 626–634 (1999)
16. C.M. Bishop, *Neural Networks for Pattern Recognition* (Oxford University Press, 2004)
17. P.M. Rice, *Pottery Analysis: A Sourcebook* (The University of Chicago Press, Chicago, 1989)
18. K.G. Harry, A. Johnson, A non destructive technique for measuring ceramic porosity using liquid nitrogen. J. Archaeological Science **31**, 1567–1575 (2004)
19. S.L. Olsen, *Scanning Electron Microscopy in Archaeology* (British Archaeological Reports, Oxford, 1998)
20. J.D. Cheeke, *Fundamentals and Applications of Ultrasonic Waves* (CRC Press LLC, USA, 2002)
21. J. Krautkrämer, *Ultrasonic Testing of Materials*, 4th edn. (Springer, Berlin, 1990)
22. M. Alba-Calzado, S. Gutiérrez-Lloret, Las producciones de transición al Mundo Islámico: el problema de la cerámica paleoandalusí (siglos VIII y IX). In Cerámicas hispanorromanas: Un estado de la cuestión, eds. B. Casasola, A. Ribera i Lacomba (Universidad de Cadiz, Spain, 2008) pp 585–613
23. A. Salazar, L. Vergara, J. Igual, J. Gosalbez, Blind source separation for classification and detection of flaws in impact-echo testing. Mech Syst Signal Process. **19**(6), 1312–1325 (2005)
24. L. Vergara, R. Miralles, J. Gosálbez, F.J. Juanes, L.G. Ullate, J.J. Anaya, M.G. Hernández, M.A.G. Izquierdo, NDE ultrasonic methods to characterize the porosity of mortar. NDT&E International, Elsevier **34**(8), 557–562 (2001)
25. A. Hyvärinen, J. Karhunen, E. Oja, *Independent Component Analysis* (John Wiley & Sons, New York, 2001)
26. A. Cichocki, S. Amari, *Adaptive Blind Signal and Image Processing: Learning algorithms and applications* (Wiley, John & Sons, New York, 2001)
27. R. Miralles, L. Vergara, J. Gosálbez, Material grain noise analysis by using higher-order statistics. Signal Process. **84**(1), 197–205 (2004)
28. J. Igual, A. Camacho, L. Vergara, Blind source separation technique for extracting sinusoidal interferences in ultrasonic non-destructive testing. J. VLSI Signal Process. **38**, 25–34 (2004)
29. J. Igual, A. Camacho, L. Vergara, Cancelling sinusoidal interferences in ultrasonic applications with a BSS algorithm for more sources than sensors, in *Proceedings of the Independent Componente Analysis Workshop*, ICA, (San Diego, 2001)
30. S. Kay, in *Spectral Estimation: Theory and Application* (Prentice-Hall, 1988)
31. L. Vergara, J. Gosalbez, J.V. Fuente, R. Miralles, I. Bosch, Measurement of cement porosity by centroid frequency profiles of ultrasonic grain noise. Signal Processing, Elsevier **84**(12), 2315–2324 (2004)
32. P.A. Regalia, in *Adaptive IIR Filtering in Signal Processing and Control*. (Marcel Dekker, 1994).
33. A.M. Pollard, C. Heron, *Archaeological Chemistry* (The Royal Society of Chemistry, Cambridge, 2008)
34. S.K. Thompson, *Sampling, Wiley-Interscience*, 2nd edn. (USA, New York, 2002)

Chapter 7
Other Applications: Sequential Dependence Modelling and Data Mining

This chapter presents two diverse applications: diagnosis of sleep disorders (apnea) and data mining in a web of a virtual campus. The first application presents a procedure to extend ICA mixture models (ICAMM) to the case of having sequential dependence in the feature observation record. We call it sequential ICAMM (SICAMM). We present the algorithm, which is essentially a sequential Bayes processor, which can be used to sequentially classify the input feature vector among a given set of possible classes. Estimates of the class-transition probabilities are used in conjunction with the classical ICAMM parameters: mixture matrices, centroids, and source probability densities. These parameters were estimated using the Mixca algorithm proposed in Chap. 3. Some simulations are presented to verify the improvement of SICAMM with respect to ICAMM. Moreover, a real data case is considered: the computation of hypnograms to help in the diagnosis of sleep disorders. Both simulated and real data analyses demonstrate the potential interest of including sequential dependence in the implementation of an ICAMM classifier.

In the second application, ICA is used as a data mining technique to discover patterns in e-learning. An ICA model is proposed that defines the sources as dimensions of the learning styles of the students. A novel non-parametric ICA and standard ICA algorithms are applied to huge historical web log data from a virtual campus in order to detect the relationship between web activities and learning styles. The data are divided by the course types into graduate courses and regular academic courses. Each of these divisions is separated into two subsets: cases with grades and cases with no grades. Web activities include events such as course access, email exchange, forum participation, news reading, chats, and achievements. Suitable learning styles of the students were positively detected for graduate courses with grades using the non-parametric Mixca algorithm.

A. Salazar, *On Statistical Pattern Recognition in Independent Component Analysis Mixture Modelling*, Springer Theses, DOI: 10.1007/978-3-642-30752-2_7,

7.1 Including Sequential Dependence in ICAMM

7.1.1 Introduction

Mixtures of independent components analyzers are progressively recognized as powerful tools for versatile modelling of arbitrary data densities [1–7]. In most cases, the final goal is to classify the observed data vector $\mathbf{x}$ (feature) in a given class from a finite set of possible classes. To this aim, the probability of every class given the observed data vector $p[C_k/\mathbf{x}]$ must to be determined. Then the class having maximum probability is selected. Bayes theorem is claimed for the practical computation of the required probabilities since it allows expressing $p[C_k/\mathbf{x}]$ in terms of the vector observation mass density. Considering K classes, we can write

$$p[C_k/\mathbf{x}] = \frac{p[\mathbf{x}/C_k]p[C_k]}{p[\mathbf{x}]} = \frac{p[\mathbf{x}/C_k]p[C_k]}{\sum_{k'=1}^{K} p[\mathbf{x}/C_{k'}]p[C_{k'}]}, \tag{7.1}$$

where the mixture model of $p[\mathbf{x}]$ is evident in the denominator of Eq. (7.1). The ICA mixture model (ICAMM) considers that the observations corresponding to a given class k are obtained by linear transformation of vectors having independent components plus a bias term: $\mathbf{x} = \mathbf{A}_k\mathbf{s}_k + \mathbf{b}_k$. Equivalently, this implies that the observation vector in a given class can be expanded around a centroid vector $\mathbf{b}_k$ in a basis formed by the columns of $\mathbf{A}_k$. It is assumed that the basis components are independent so that the $\mathbf{A}_k$ matrix is nonsingular. When this assumption becomes invalid, due, for example, to a high dimension of the observation vector, some dimension reduction techniques like classical PCA are routinely used. The transforming matrix, the centroid, and the marginal probability density functions (which can be arbitrary) of the independent components of $\mathbf{s}_k$ (called sources) define a particular class.

Using standard results from probability theory, we have that $p[\mathbf{x}/C_k] = |\det \mathbf{A}_k^{-1}|p[\mathbf{s}_k]$. On the other hand, algorithms for learning the ICAMM parameters ($\mathbf{A}_k$, $\mathbf{b}_k$, $p[\mathbf{s}_k]\; k = 1\ldots K$) in supervised or unsupervised frameworks can be found in the given references [1–7]. Therefore, if the classifier has been trained, we can compute the required probabilities using

$$p[C_k/\mathbf{x}] = \frac{|\det \mathbf{A}_k^{-1}|p[\mathbf{s}_k]p[C_k]}{\sum_{k'=1}^{k} |\det \mathbf{A}_{k'}^{-1}|p[\mathbf{s}_{k'}]p[C_{k'}]} \quad \mathbf{s}_k = \mathbf{A}_k^{-1}(\mathbf{x} - \mathbf{b}_k). \tag{7.2}$$

However, the classes and observations very often do not appear in a totally random manner, and they exhibit some degree of sequential dependence in time or space domains. This means that the computation of the class probabilities should consider the whole history of observations. Thus, if we define the indexed matrix of observations $\mathbf{X}(n) \equiv [\mathbf{x}(0)\,\mathbf{x}(1)\ldots\mathbf{x}(n)]$, we should compute $p[C_k(n)/\mathbf{X}(n)]$.

In Sect. 7.1.2, we present a procedure to extend ICAMM to the case of sequential dependence (SICAMM). In Sect. 7.1.3, some experiments are included

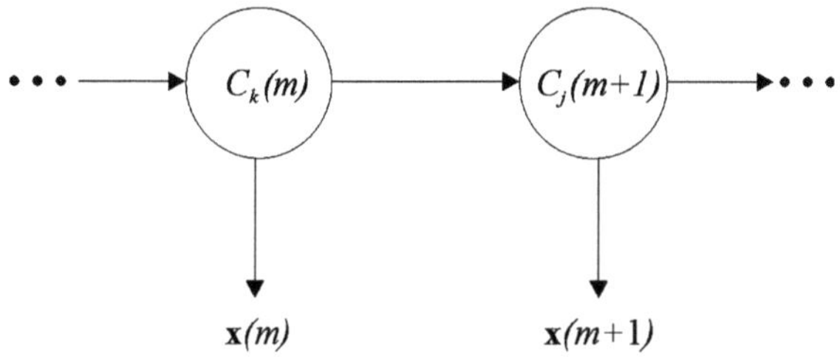

Fig. 7.1 HMM description of the class-conditionally independence between successive observation vectors

with simulated and real data showing that the classification error percentage can be reduced by SICAMM in comparison with ICAMM.

7.1.2 Sequential ICAMM

To compute $p[C_k(n)/\mathbf{X}(n)]$, we start from Eq. (7.1). We assume that, conditional on $C_k(m)$, the observed vectors $\mathbf{x}(m)\mathbf{m} = \mathbf{0}\ldots n$ are independent. This is a key assumption in the classical Hidden Markov Model (HMM) [8] structure that is described in Fig. 7.1. Statistical dependences between two successive instants are defined by the arrows connecting the successive classes. However, successive observed vectors are not directly connected, i.e., the distribution of every $\mathbf{x}(m)$ is totally defined if we know the corresponding class $C_k(m)$. In particular, this implies that $p[\mathbf{x}(n)/C_k(n)] = p[\mathbf{x}(n)/C_k(n)] \cdot p[\mathbf{x}(n-1)/C_k(n)]$. We developed the details of the SICAMM algorithm in [9].

Let us describe the SICAMM algorithm in a more specific form. We assume that the parameters $\mathbf{A}_k$, $\mathbf{b}_k$, $p[\mathbf{s}_k]\, k = 1\ldots K$ have been previously estimated by means of an ICAMM learning algorithm from the several algorithms available in the literature and that the class-transition probabilities are also known or estimated. Table 7.1 describes the algorithm.

Note that the SICAMM algorithm can be expressed in the form of a sequential Bayesian processor [8]

$$p[C_k(n)/\mathbf{X}(n)] = W_k(n) \cdot p[C_k(n)/\mathbf{X}(n-1)] \quad W_k(n) = \frac{p[\mathbf{x}(n)/C_k(n)]}{p[\mathbf{x}(n)/\mathbf{X}(n-1)]}, \tag{7.3}$$

where $p[C_k(n)/\mathbf{X}(n-1)]$ is a “prediction” of the current class given the past history of observations and where $W_k(n)$ is an “updating weight” that measures the significance of the current class relative to the significance of the past history of observations for generating the current observation.

Table 7.1 SICAMM algorithm

Initialization $n = 0$

$\mathbf{X}(0) = [\mathbf{x}(0)]$

$\mathbf{s}_k(0) = \mathbf{A}_k^{-1}(\mathbf{x}(0) - \mathbf{b}_k)\ k = 1\ldots K;$

$$p[C_k(0)/\mathbf{X}(0)] = \frac{\left|\det \mathbf{A}_k^{-1}\right| p[\mathbf{s}_k(0)]}{\sum_{k'=1}^{K} \left|\det \mathbf{A}_{k'}^{-1}\right| p[\mathbf{s}_{k'}(0)]}$$

For $n = 1$ **to** N

$\mathbf{X}(n) = [\mathbf{x}(0)\,\mathbf{x}(1)\ldots\mathbf{x}(n)]$

$\mathbf{s}_k(n) = \mathbf{A}_k^{-1}(\mathbf{x}(n) - \mathbf{b}_k) \quad k = 1\ldots K$

$$p[C_k(n)/\mathbf{X}(n-1)] = \sum_{k'=1}^{K} p[C_k(n)/C_{k'}(n-1)] \cdot p[C_{k'}(n-1)/\mathbf{X}(n-1)]$$

$$p[C_k(n)/\mathbf{X}(n)] = \frac{\left|\det \mathbf{A}_k^{-1}\right| p[\mathbf{s}_k(n)] p[C_k(n)/\mathbf{X}(n-1)]}{\sum_{k'=1}^{K} \left|\det \mathbf{A}_{k'}^{-1}\right| p[\mathbf{s}_{k'}(n)] p[C_{k'}(n)/\mathbf{X}(n-1)]}$$

7.1.3 Simulations

We have considered a simple scenario that is similar to the first example included in the classical ICAMM [1]. Observations are vectors of dimension 2, and the number of classes is also 2. In class 1, the observation vectors are obtained by linearly transforming independent component vectors where both components are obtained from uniform distributions having zero mean and unit variance. In class 2, the observation vectors are obtained in the same way as in class 1, but the distributions are zero mean and unit variance Laplacian. The centroids were selected relatively close, thus $\mathbf{b}_1 = [1\ 1]^T$ and $\mathbf{b}_2 = [1.5\ 1.5]^T$. We have compared the error percentages in classifying an observation as belonging to class 1 or to class 2 using ICAMM and SICAMM. The parameters of the ICA mixtures were estimated using the Mixca procedure of Chap. 3 for both simulations and for the application of analysis of hypnograms.

First of all, to simplify the comparison, we have considered that the probability of staying in the same class is the same for both class 1 and class 2. Therefore, only one parameter α is required to establish the degree of sequential dependence, i.e.,

$$\begin{aligned} p[C_1(n)/C_1(n-1)] &= p[C_2(n)/C_2(n-1) = \alpha] \\ p[C_1(n)/C_2(n-1)] &= p[C_2(n)/C_1(n-1)] = 1-\alpha \end{aligned} \tag{7.4}$$

In Fig. 7.2, we represent the estimated error percentages for α varying from 0.5 (no sequential dependence at all) to 1 (total dependence).

The error percentage was estimated by the quotient between the number of misclassified observations and the total number of generated observations. To obtain reliable results we considered 300 pairs of transforming matrices, $\mathbf{A}_1$ and $\mathbf{A}_2$, which were randomly generated: every element of these matrices was obtained from a uniform distribution between 0 and 1. For every pair of matrices, a run of 200 observation vectors was produced taking the above defined centroids and distributions into account. Hence, the total number of observation vectors for every α value was 60,000.

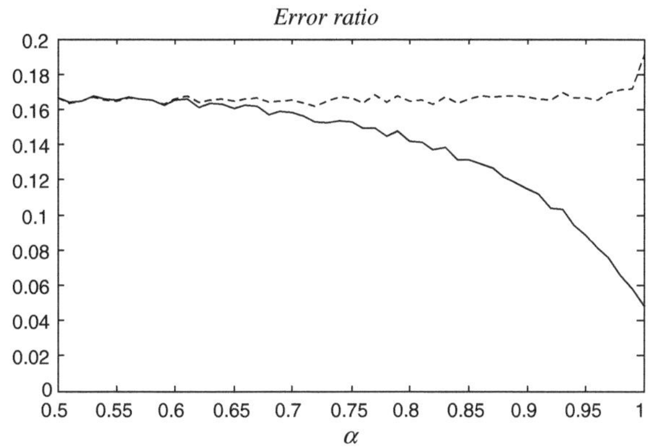

Fig. 7.2 Average error ratios in the classification between class 1 and class 2 for different values of the class persistence parameter α, corresponding to SICAMM (*solid line*) and ICAMM (*dotted line*)

Note that for $\alpha > 0.6$ the error percentage of classification starts to decrease when using SICAMM, while it remains constant when using ICAMM. This is an obvious consequence of the capability of SICAMM to exploit the sequential dependence of successive classes.

7.1.4 Analysis of Hypnograms

In this section, we show a practical application of SICAMM to a real data problem. The framework is computer-assisted sleep staging [10]. Typically, human sleep is divided into four different stages: awake, light sleep, deep sleep, and rapid-eye movement (REM) sleep. Determining the stage of a patient over a long period is relevant to the diagnosis of different sleep disorders and other human illnesses. The sequential record of the different sleep stages corresponding to a given period is called a hypnogram. Hypnograms of a patient are usually obtained in an non-automatic manner by experts making a visual inspection of the so-called polysomnograms (PSG), which is a set of EEGs and other records obtained from the patient while sleeping. Some degree of automaticity has been implemented to help the experts [10], but a totally automatic system is still a challenge.

Of particular interest is the detection of very short periods of wakefulness [11] (also called arousals) since their frequency of appearance are related to the presence of apnea and epilepsy. Thus, we have applied ICAMM and SICAMM to obtain the automatic detection of arousals. Hence, the hypnograms will show only two stages: "stage 1", which corresponds to sleep (in any of the possible sleep stages); and "stage 2", which corresponds to arousal. The results obtained from ICAMM and SICAMM are compared with the non-automatic detection made by an expert.

The observation vector, which is the input to the classifier, has four feature elements. Every feature is computed in very short segments of the PSG signals (typically 1–3 s) and averaged in epochs of 30 s. Then, a decision about the sleep

Table 7.2 Main data corresponding to the sleep staging experiment

Patient	Testing sample size (number of epochs)	Training sample size (number of epochs)	Estimated stage transition probabilities	Percentage error ICAMM (%)	Percentage error SICAMM (%)
1	445 (3.7 h) 424 "stage 1" 21 "stage 2"	445 (3.7 h) 402 "stage 1" 43 "stage 2"	P11 = 0.97 P22 = 0.69	20	9
2	454 (3.8 h) 403 "stage 1" 51 "stage 2"	454 (3.8 h) 361 "stage 1" 93 "stage 2"	P11 = 0.96 P22 = 0.80	36	20

stage is made every 30 s. The selected four features extracted from the PSG signals were: the amplitude, the dominant rhythm, and the theta-slow-wave index (TSI), which were estimated from the C3-A2 EEG channel, and the alpha-slow-wave index (ASI), which was estimated from the O2-A1 EEG channel. The dominant rhythm was estimated as the pole frequency of the second-order autoregressive (AR) model; the ASI was the ratio of power in the alpha band (8.0–11 Hz) to the combined power in the delta (0.5–3.5 Hz) and theta (3.5–8.0 Hz) bands; and the TSI was the ratio of power in the theta band to the combined power in the delta and alpha bands. These features are commonly used in computerized PSG analysis [10, 11].

Two patients with apnea were considered for the experiment. The main data are included in Table 7.2. The parameters $\mathbf{A}_k$, $\mathbf{b}_k$, $p[\mathbf{s}_k]\ k = 1\ldots K$ were estimated from the training record in a supervised form using the JADE algorithm [12]. The labelling of stages "1" and "2" that is required for the supervised training was made taking into account the manual score done by the expert. The probabilities of transition between stages were also estimated from the training record. Note that the probabilities of permanence in the same class are clearly above 0.5, so the use of SICAMM seems to be justified in this application. The reference hypnograms were also obtained by an expert using conventional non-automatic procedures. Obviously, there is a clear improvement of SICAMM with respect to ICAMM when we compare the percentage of error in the automatic computation of the hypnograms with respect to the reference hypnogram.

To have a better understanding of the results, in Figs. 7.3 (patient 1) and 7.4 (patient 2), we show the hypnogram estimated by the expert, together with the hypnograms computed by SICAMM and ICAMM. Essentially, SICAMM reduces the number of false detections of arousals so that a "cleaner" hypnogram is obtained. In patient 1, we verified that 9 of the misclassified stages obtained with SICAMM actually corresponded to "stage 2" and 31 corresponded to "stage 1". With ICAMM, 9 of the misclassified stages obtained actually corresponded to "stage 2" and 81 corresponded to "stage 1". Similarly, in patient 2, we verified that 31 of the misclassified stages obtained with SICAMM actually corresponded to "stage 2" and 61 corresponded to "stage 1". With ICAMM, 31 of the misclassified stages obtained actually corresponded to "stage 2" and 165 corresponded to "stage 1".

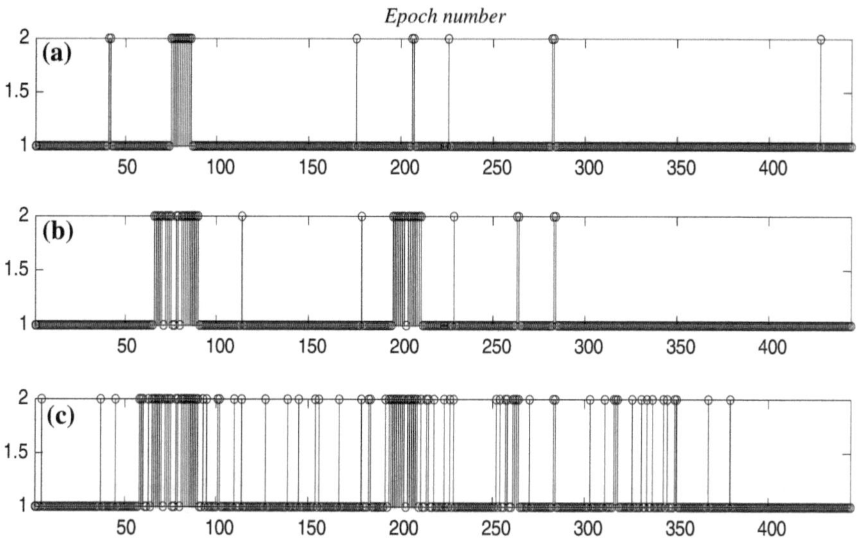

Fig. 7.3 Hypnograms corresponding to patient 1. **a** Expert hypnogram. **b** SICAMM hypnogram. **c** ICAMM hypnogram

7.2 Webmining Application for Detecting Learning Styles in Virtual Education

7.2.1 Introduction

Pattern analysis or data mining from web log data (webmining) is a new research area that attempts to understand the flow of information on the web by means of automated techniques for knowledge search 0 [13]. This area has a wide range of emergent applications including e-learning, e-commerce, automated information assistants, and many applications that operate through the web [14]. One example of webmining is the classification of web pages based on understanding the textual content of emails based on hierarchical probabilistic clustering [15].

Specifically, there is an increasing interest in webmining from e-learning data. Some examples are: predicting the drop-out rate using demographic (sex, age, marital status, etc.) and course data [16]; predicting the course grades using the processing success rate, success at first try, number of attempts, time spent on the problem, etc. [17]; combining several weak classifiers by boosting in order to predict the final grades [18]. Recently, new holistic webmining approaches have undertaken the extraction of learning styles from web navigational behaviour outlined by students [19–21].

The approaches used to extract patterns of the web log data have used different statistical classification and machine learning techniques including ICA. As explained in Chap. 2, many of the standard ICA algorithms are based on a

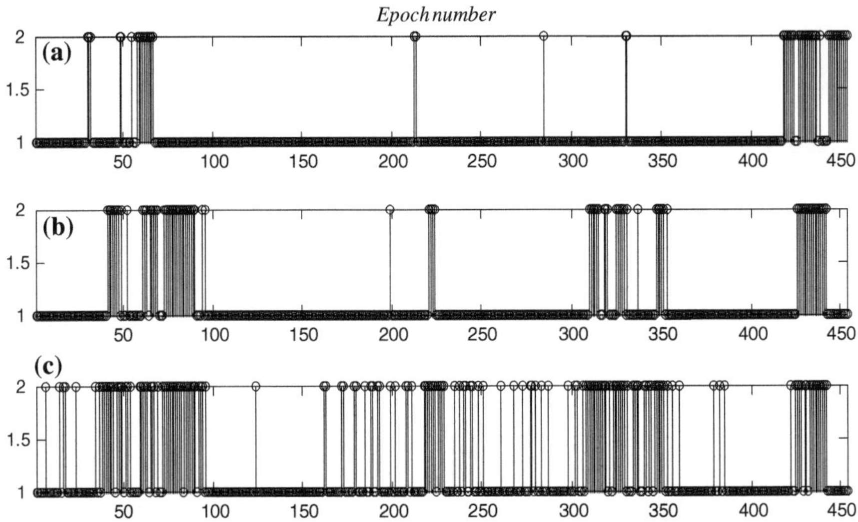

Fig. 7.4 Hypnograms corresponding to patient 2. **a** Expert hypnogram. **b** SICAMM hypnogram. **c** ICAMM hypnogram

parametric model of the source pdf; however there are recent contributions based on non-parametric density estimation of the sources (see for instance [22, 23]).

In this section, we provide an application of non-parametric ICA to detect learning styles in e-learning. This was carried out on data of graduate and undergraduate courses at the *Universidad Politécnica Abierta* (UPA) site. The UPA is a virtual campus at the *Universidad Politécnica de Valencia*, which at the time the data was collected in 2005 had more than 6,000 students registered in about 230 courses. Figure 7.5 shows a general schema of the facilities at the virtual campus learning environment. The e-learning event activities at the campus web were analyzed to recognize patterns on learning styles of the students.

Data from the use of the UPA web facilities included the following information about e-learning event activities: 1(course access), 2(agenda using), 3(news reading), 4(content consulting), 5(email exchange), 6(chats), 7(workgroup document), 8(exercise practice), 9(course achievement), and 10(forum participation). The date and time for each event were also available. Besides the information on the web activity, the exercises performed and the grades obtained by the UPA students were also available. The data were collected from the virtual campus web in the period from January 2002 to March 2005, totalling 2,391,003 records.

A learning-style model classifies students according to where they fit on a number of scales corresponding to the ways in which they receive and process information. One of the most accepted learning style taxonomies for engineering students is Felder's model [24]; see Table 7.3 (one learning style is formed by the combination of one feature in each dimension, for instance, intuitive-visual-deductive-active-global). This model was used in the present work.

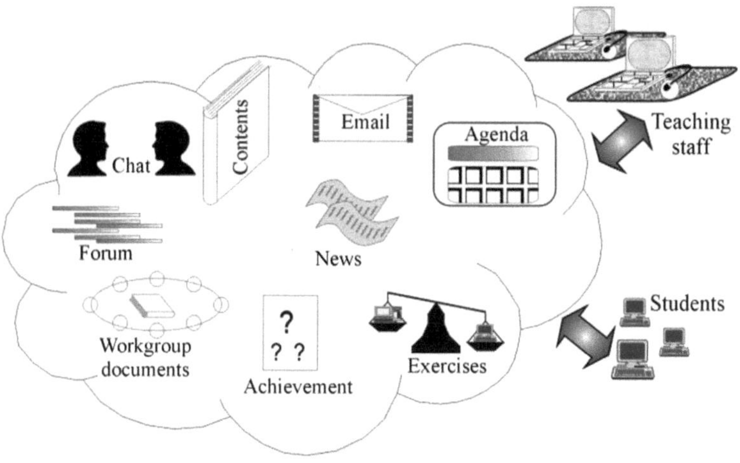

Fig. 7.5 E-learning facilities outline at UPA

Table 7.3 Dimensions of learning and teaching styles (Felder's model)

	Preferred learning style		Corresponding teaching style	
1	Sensory–intuitive	Perception	Concrete–abstract	Content
2	Visual–auditory	Input	Visual–verbal	Presentation
3	Inductive–deductive	Organization	Inductive–deductive	Organization
4	Active–reflective	Processing	Active–passive	Student participation
5	Sequential–global	Understanding	Sequential–global	Perspective

7.2.2 ICA Statement of the Problem

In order to solve the problem of detecting learning styles in e-learning, we defined the following ICA generative model for the observed data.

$$\mathbf{x} = \mathbf{E} \times \mathbf{d} \tag{7.5}$$

We assume that the underlying independent sources that generate the web log data are dimensions of the learning styles of the students, and we observe **x** linear combinations of those styles through the use of the facilities at the virtual campus. Then, d_i, $(i = 1, \ldots, 5\ \textit{learning style dimension})$ corresponds to the "perception", "input", "organization", "processing", and "understanding" dimensions (see Table 7.3); and the mixture matrix **E** provides the relation between the e-learning style dimensions and the e-learning event activities, $\mathbf{e}_{ij}$, $(i = 1, \ldots, 5\ \textit{learning style dimension})$, $(j = 1, \ldots, 10\ e\text{–learning activity})$.

Some preprocessing was done; a datawarehouse was created from the historical (2001–2005) web data of the UPA. The total number of events for the analyzed period was 2,391,003 (see the list of e-learning event activities in Sect. 7.2.1). The projection of this event table on the student codes was a table with 8,909 records.

For each student, the corresponding total instance counter of each kind of event was calculated, and a normalized value (1–100 scale) of student event activity was calculated as

$$even\_activity_{student} = \frac{event\ total\ instance_{student}}{instance\ maximum_{event}} \cdot 100 \tag{7.6}$$

The student activity data were added as fields to the datawarehouse plus the average connection time to the web and average grade obtained for each student. Course achievement was not always required in virtual courses at UPA, so only 1,873 of the 8,909 rows of the datawarehouse had a value for an average grade. Besides, those data subsets were divided according to whether they were graduate or regular academic courses.

Once the data were prepared, we applied the non-parametric Mixca algorithm, which was described in Chap. 3, configuring it to estimate one ICA (see Appendix A). The results are presented in the next section.

7.2.3 Results and Discussion

The ICA algorithm was applied in two stages. In the first stage, ICA was applied directly to the data in order to determine whether the e-learning web activities were independent by themselves, i.e., searching for those web activities that can be separated by an ICA algorithm as a source. The second stage consisted of applying ICA on a reduced data dimension of five components in order to associate those components with Felder's learning dimensions.

Figure 7.6 shows the data of the web activities and average connection time and grades for graduate course with grades. Figure 7.7 shows the sources recovered by non-parametric ICA of those data. It is a high correlation between event 7 (workgroup documents) and the source s9, and between event 8 (exercise practice) and the source s5. Therefore, we can assume independence for those events.

After analyzing the results from the ICA applied to the different data subsets, we can infer the following conclusions:

- Email exchange was independent in some cases. In some courses, e-mail exchange was not mandatory in the activities.
- The workgroup document event was independent. In the case of courses with no grades, the lack of evaluation and grades discouraged the participation of students in collaborative tasks. In the cases of courses with grades, it was an optional activity in some courses.
- In some datasets, the content consulting event was independent as a reflection of the distributed passive learning (DPL) nature of the web platform [25]. Thus, content consulting became a routine consisting in download materials with no interactive learning process.

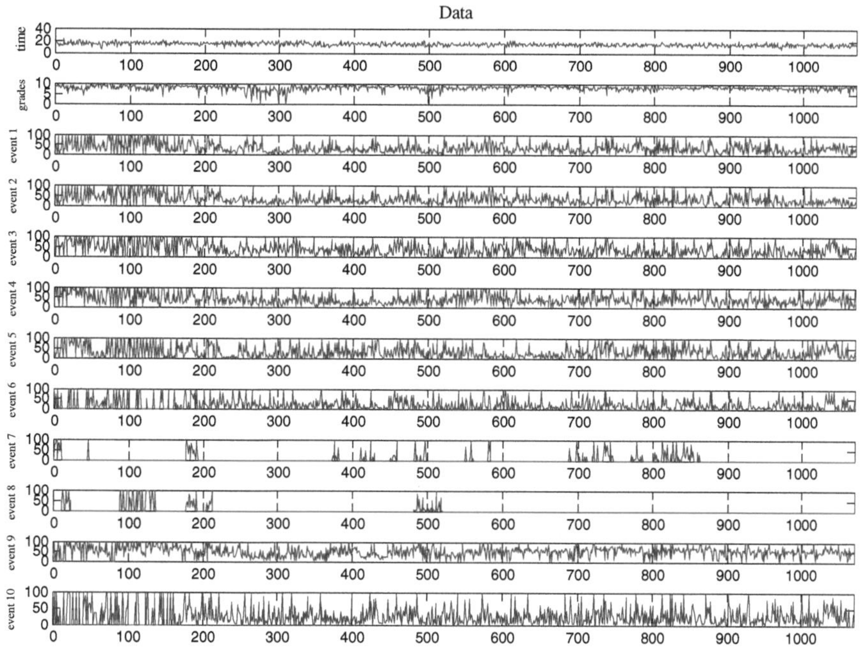

Fig. 7.6 Data of the graduate courses with grades: event 1 (course access) ... event 10 (forum participation)

- Exercise practice and course achievement were also found to be independent events for some datasets. This could be due to the profile of some students, which included a background in information and telecommunications technology as well as knowledge about the course content. For those students, participating in those event activities would be irrelevant.

In a second stage, we applied ICA after reducing the data to five components by PCA [26]. We applied PCA to group the events of the web activity into learning dimensions taking into account Felder's framework [24]. PCA reduced 10 web event activities to five components. Table 7.4 contains the first six sorted contributions of the web activities of the ICA mixture matrix for the five estimated sources. Each source was associated with one learning dimension from Table 7.3, analyzing the weight of the web activities and considering the principal evaluation methodologies employed by teachers for graduate courses with grades. -Dimension 1 was not detected and Dimension 5 was detected twice (5 and 5')-. The methodologies assigned grades focusing on: achievement, individual student participation, or group work. The implicit teaching styles of the evaluation methodologies encouraged specific learning styles of the students, as we explain below.

The learning dimension 1 (sensory–intuitive) corresponding to "perception" was not detected in the ICA mixing matrix. This could be due to the fact that the

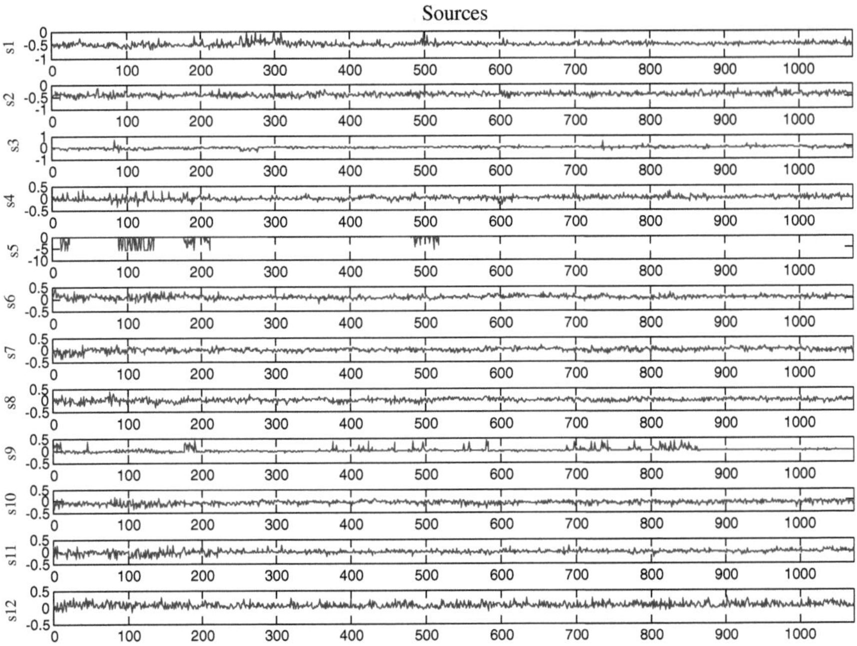

Fig. 7.7 Sources calculated by the Mixca algorithm for the data of the graduate courses with grades

Table 7.4 ICA mixing matrix

LSD[a]	Sorted web activity contribution					
2	Chat	Forum	News	Email	Access	Exercises
	1	0.82283	0.30755	0.16476	0.14756	0.14231
4	Email	Content	Wg-doc[b]	Exercises	Forum	Chat
	1	0.34189	0.32297	0.28768	0.22548	0.20078
3	Wg-doc	News	Achieve	Content	Chat	Email
	1	0.80531	0.4122	0.39987	0.39421	0.31666
5'	Achieve	Content	Agenda	Access	Forum	News
	1	0.45124	0.2117	0.21116	0.20087	0.18239
5	Access	Agenda	Content	Achieve	Email	Chat
	1	0.95776	0.85549	0.7143	0.5832	0.49774

[a] Learning style dimension
[b] Workgroup documents

emphasis of educational strategies did not favour highlighting that dimension. The relationship between learning style dimensions and web activities can be made using Table 7.4. Table 7.5 shows the association detected by ICA between learning styles and web activities (we have added a possible web activity combination for learning dimension 1).

Table 7.5 Association between learning styles and web activities

	Learning style		Web event activity
1	Sensory–intuitive	Perception	Chats, forum participation, course access
2	Visual–auditory	Input	Chats, forum participation, news reading, email exchange
3	Inductive–deductive	Organization	Workgroup document, news reading, course achievement, content consulting
4	Active–reflective	Processing	Email exchange, content consulting, workgroup document, exercise practice
5	Sequential–global	Understanding	Course access, agenda using, content consulting, course achievement

Note that some web activities are associated with more than one dimension; this makes sense because a web activity could require several capabilities of the students to be used in their learning process. By allowing this kind of relationship, we can obtain more real and versatile descriptions of student learning styles as well as include all the dimensions of the learning framework. In [21], just three dimensions of Felder's model were considered and the Bayesian network proposed constrained the relationship of the web activities with just one dimension of the learning model.

Figure 7.8 shows the sources 3, 4, and 5 (organization, processing, understanding) obtained for the graded graduate course dataset. Four labelled characterised zones in the learning style space are displayed: (1) This zone represents the most important learning style in the population. The learning for the students in this zone emphasizes global understanding, active processing, and deductive logic (natural human teaching style), and high grades. (2) This zone focuses on inductive logic (natural human learning style), sequential understanding, and relative active processing. Students within this learning style could have natural skills for virtual education. (3) This zone is characterised by global understanding, deductive logic, and reflective processing. Students within this learning style would have higher abstraction skills that require teaching. (4) This zone basically represents outliers with individual learning styles.

We can conclude that the understanding dimension facilitates the clear projection of the learning styles of the students. Its principal components are course achievement, content consulting, and use of agenda. This finding confirms the assumption that the quickest way to change the learning style of the student is to change the assessment style, i.e., expected evaluation biases how the student learns [27].

We made a cluster validation procedure to determine the best quality of cluster configuration for the data in Fig. 7.8. It consisted of estimating the partition and partition entropy coefficients for different number of clusters [28], which were explained in the Introduction chapter. The evolution of those coefficients is shown in Fig. 7.9. By applying the criterion to Fig. 7.9 to detect the best partitioning (the point where the slope of the two curves changes), we found that the optimum number of clusters is 4.

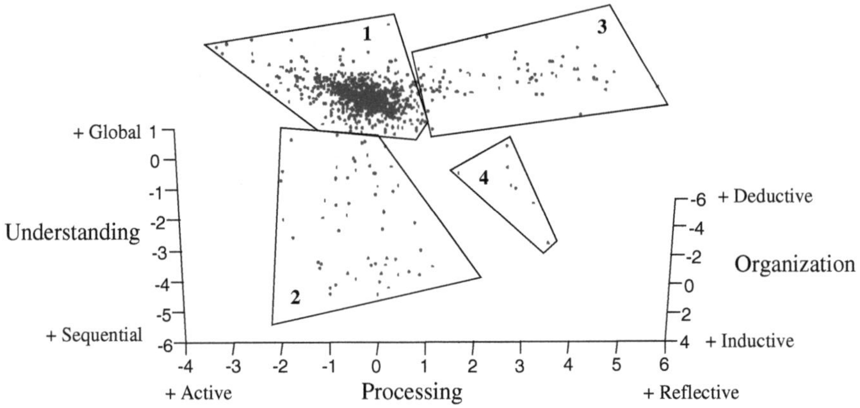

Fig. 7.8 Three sources in a learning style space for graduate courses with grades

Fig. 7.9 Measures of clustering validation with three sources

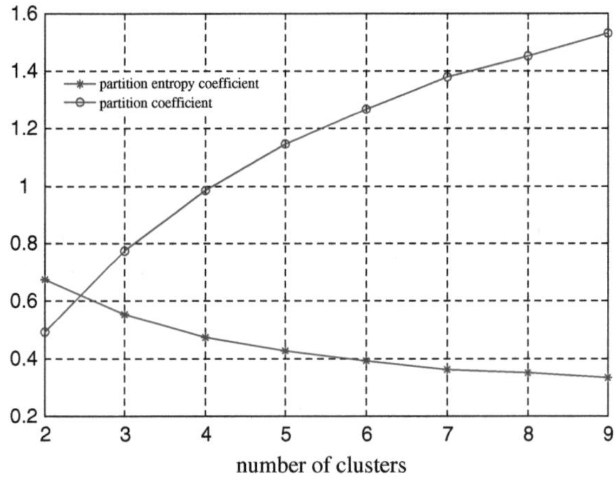

Figure 7.10 shows three sources for graduate courses with no grades. The distribution of the data in Fig. 7.10 does not allow learning style groups to be formed and shows all the subjects within a single learning style. Since the understanding and organization dimensions do not discriminate the projection of the learning styles, only the processing dimension provides some discrimination. Therefore, a single learning style that emphasizes reflection over practice is obtained. This is apparent in the content consulting and exercise practice components of the processing dimension. The conclusion is that the lack of assessment does not allow student learning styles to be developed.

The results for regular academic courses were similar to the graduate course results finding meaningful learning styles for courses with grades. Non-parametric ICA provided better results than standard ICA algorithms allowing more suitable learning styles clusters to be detected. Classical ICA algorithms rely on

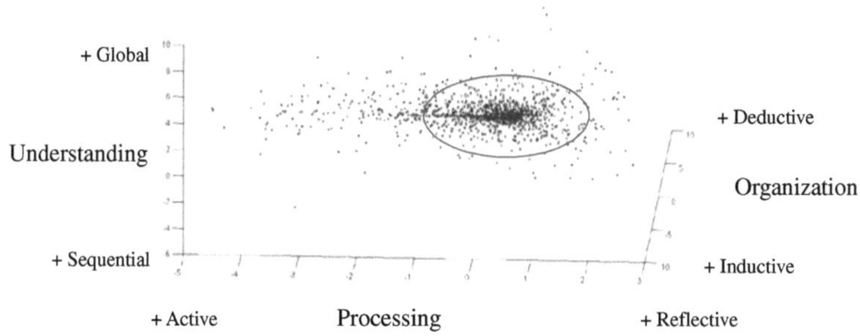

Fig. 7.10 Three sources in a learning style space for graduate courses with no grades

assumptions about the source signals so that they imply a given model for the source distributions, for instance, the Extended InfoMax algorithm [29] restricts the sources to be sub-gaussian or super-gaussian, or also classical ICA rely exclusively on certain higher order statistics of the signals in order to measure independence, such as [30]. Other algorithms make assumptions that only fit to specific applications like TDSEP [31], which is based on some time-delayed statistics and is not suited for problems without temporal correlations. The non-parametric ICA algorithm applied is more adaptable to the data. It does not assume any restriction on the data since the probability distributions are calculated directly from the training set through a non-parametric kernel-based approach. This algorithm also focuses the independence estimation among the source components directly based on the marginal distributions.

Summarizing the results, Fig. 7.11 shows the relationship between learning style dimensions and e-learning web activities separating interactivity from personal activity fields.

7.3 Conclusions

In this chapter, we have explored the use of the ICAMM and the SICAMM methods in two quite different applications: sequential dependence modelling and data mining. The parameters of the ICA mixture models were estimated using the Mixca algorithm proposed in Chap. 3. In the first application, we proposed an extension of ICAMM, which allows class-transition information to be included in the classifier. As this is equivalent to considering the sequencing of the classes, we have called it sequential ICAMM. Essentially, SICAMM is a sequential Bayesian processor where the underlying probability densities are mixtures of independent component analyzers. Estimates of the model parameters (mixing matrices, centroids, and probability densities of the sources) are required in both ICAMM and SICAMM. This can be done in a supervised (true classes are known) or unsupervised (true classes are estimated) manner from a training set of feature vectors.

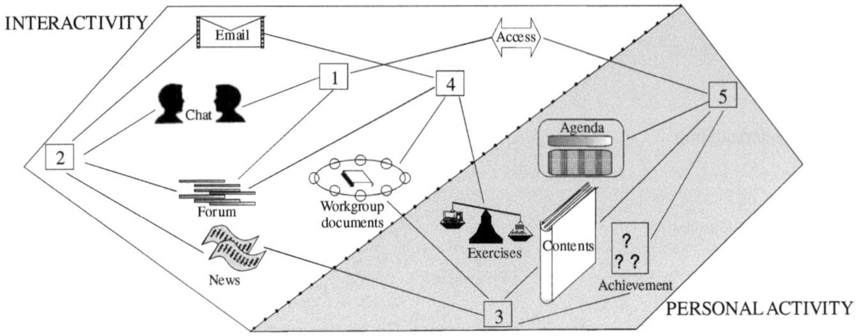

Fig. 7.11 Outline of learning style dimensions associated with e-learning web activities obtained by non-parametric ICA

Once the true classes (known or estimated) are available, the class-transition probabilities can be easily estimated.

Some simulations and a real data analysis case have verified the potential improvements derived from including sequential information in the ICAMM algorithms that are used for classification purposes. Further work is required to develop algorithms which can simultaneously estimate all the model parameters in an unsupervised framework.

The second application, which consisted of applying the proposed non-parametric ICA to detect the patterns of learning styles in educational web activities, produced promising results for a real case with huge historical data. Modelling learning dimensions as a combination of web event activities enhanced the detection of the student learning styles. Those results could be used to adapt teaching methodologies or, in general, to improve the learning system, balancing distributed passive learning (DPL) and distributed interactive learning (DIL).

Thus, the versatility of these methods has been demonstrated into different kinds of problems; the first problem where there are hidden variables that model dynamic dependence among class transitions; and the second problem, where the proposed non-parametric approach is able to estimate suitable sources for the analysis of huge data (from which it is normally difficult to derive a parametric model for the source distributions).

References

1. T.W. Lee, M.S. Lewicki, T.J. Sejnowski, ICA mixture models for unsupervised classification of non-gaussian classes and automatic context switching in blind signal separation. IEEE Trans. Pattern Anal. Mach. Intell. **22**(10), 1078–1089 (2000)
2. T.W. Lee, M.S. Lewicki, Unsupervised image classification, segmentation, and enhancement using ICA mixture models. IEEE Trans. Image Process. **11**(3), 270–279 (2002)
3. R. Choudrey, S. Roberts, Variational mixture of bayesian independent component analysers. Neural Comput. **15**(1), 213–252 (2002)

4. N.H. Mollah, M. Minami, S. Eguchi, Exploring latent structure of mixture ICA models by the Minimum ß-Divergence method. Neural Comput. **18**, 166–190 (2005)
5. C.T. Lin, W.C. Cheng, S.F. Liang, An on-line ICA-mixture-model-based self-constructing fuzzy neural network. IEEE Trans. Circuits Syst. **52**(1), 207–221 (2005)
6. C.A. Shah, P.K. Varshney, M.K. Arora, ICA mixture model algorithm for unsupervised classification of remote sensing imagery. Int. J. Remote Sens. **28**(8), 1711–1731 (2007)
7. A. Salazar, L. Vergara, A. Serrano, J. Igual, A general procedure for learning mixtures of independent component analyzers. Pattern Recognit. **43**(1), 69–85 (2010)
8. O. Cappe, E. Moulines, T. Ryden, *Inference in Hidden Markov Models* (Springer, New York, 2005)
9. A. Salazar, L. Vergara, R. Miralles, On including sequential dependence in ICA mixture models. Signal Process. **90**, 2314–2318 (2010)
10. R. Agarwal, J. Gotman, Computer-assisted sleep staging. IEEE Trans. Biomed. Eng. **12**(48), 1412–1423 (2001)
11. M. Jobert, H. Shulz, P. Jähnig, C. Tismer, F. Bes, H. Escola, A computerized method for detecting episodes of wakefulness during sleep based on the Alpha slow-wave index (ASI). Sleep **17**(1), 37–46 (1994)
12. J.F. Cardoso, A. Souloumiac, Blind beamforming for non gaussian signals. IEE Proc.-F **140**(6), 362–370 (1993)
13. A.K. Jain, Statistical pattern recognition: a review. IEEE Trans. Pattern Anal. Mach. Intell. **22**(1), 4–37 (2000)
14. J. Srivastava, R. Cooley, M. Deshpande, P. Tan, Web usage mining: discovery and applications of usage patterns from web data. SIGKDD Explor. **2**(1), 12–23 (2000)
15. J. Larsen, L.K. Hansen, A. Szymkowiak, T. Christiansen, T. Kolenda, Webmining: learning from the World Wide Web, special issue of Computational Statistics and Data Analysis. Comput. Stat. Data Anal. **38**, 517–532 (2002)
16. S.B. Kotsiantis, C.J. Pierrakeas, P.E. Pintelas, Preventing student dropout in distance learning using machine learning techniques. *Proceedings of 7th International Conference on Knowledge-Base Intelligent Information an Engineering Systems*, pp. 267–274 (2003)
17. B. Minaei, D.A. Kashy, G. Kortemeyer, W. Punch, Predicting student performance: an application of data mining methods with an educational web-based system. *Proceedings of 33rd Frontiers in Education Conference*, pp. T2A-13-T2A-18 (2003)
18. W.Zang, F. Lin, Investigation of web-based teaching and learning by boosting algorithms. *IEEE International Conference on Information Technology: Research and Education*, pp. 445–449 (2003)
19. E. Mor, J. Minguillón, E-learning personalization based on itineraries and long-term navigational behavior. *Proceedings of 30th World Web Conference,* no. 2, pp. 264–265, New York, 2004
20. M. Xenos, Prediction and assessment of student behaviour in open and distance education in computers using Bayesian networks. Comput. Education **43**, 345–359 (2004)
21. P. Garcia, A. Amandi, S. Schiaffino, M. Campo, Evaluating Bayesian Networks' Precision for Detecting Students' Learning Styles. Comput. Education **49**, 794–808 (2007)
22. R. Boscolo, H. Pan, Independent component analysis based on nonparametric density estimation. IEEE Trans. Neural Netw. **15**(1), 55–65 (2004)
23. E.G. Learned-Miller, J.W. Fisher, ICA using spacings estimates of entropy. J. Mach. Learn. Res. **4**, 1271–1295 (2003)
24. R. Felder, L. Silverman, Learning and teaching styles. J. Eng. Education **78**(7), 674–681 (1988)
25. M. Khalifa, R. Lam, Web-based learning: effects on learning process and outcome. IEEE Trans. Education **45**(4), 350–356 (2002)
26. W. Hardle, L. Simar, *Applied Multivariate Statistical Analysis* (Springer, New York, 2006)
27. L.R.B. Elton, D.M. Laurillard, Trends in research on student learning. Stud. High. Education **4**, 87–102 (1979)

28. M. Haldiki, Y. Batistakis, M. Vazirgiannis, On clustering validation techniques. J. Intell. Inf. Syst. **17**(2–3), 107–145 (2001)
29. T.W. Lee, M. Girolami, T.J. Sejnowski, Independent component analysis using an extended InfoMax algorithm for mixed sub-gaussian and super-gaussian sources. Neural Comput. **11**(2), 417–441 (1999)
30. J.F. Cardoso, High-order contrasts for independent component analysis. Neural Comput. **11**(1), 157–192 (1999)
31. A. Ziehe, K.R. Müller, TDSEP—an efficient algorithm for blind separation using time structure. *Proceedings of the 8th International Conference on Artificial Neural Networks, ICANN'98, Perspectives in Neural Computing*, pp. 675–680 (1998)

Chapter 8
Conclusions

The overall objective of this thesis was to research pattern recognition based on the modelling of the M-dimensional probability density function of the data using mixtures of independent component analyzers. The proposed methods define a general framework that is suitable for application to problems that involve complex probability densities. In order to evaluate the potential of the proposed methods, novel applications in several fields were explored. Therefore, the capabilities of the methods to solve real-world problems has been demonstrated.

This chapter summarizes the research findings, revisiting the specific objectives given in the Introduction chapter. Section 8.1 reviews the contents of this work, drawing out the main conclusions that were derived from each chapter. The contributions of this dissertation are included in Sect. 8.2. Recommendations for future research lines are listed in Sect. 8.3.

8.1 Summary

The first two chapters presented the motivation, problems, and techniques sought in the thesis. The outlined problems focused on signal classification and blind source separation (BSS). Thus, the fundamental area of research to deal with these problems was independent component analysis (ICA) and its extension to mixtures of ICA models (ICAMM). Two principal methods for classification and hierarchical clustering which incorporate unsupervised, semi-supervised, and supervised learning were proposed. These methods were evaluated in diverse applications in order to solve real-world problems.

ICAMM has established a framework for non-linear processing of data with complex non-gaussian distributions. Data complexity is captured by a combination of local linear ICA projections and, thus, the resulting probability density function of the final model can be used to model class-conditional densities. In addition,

A. Salazar, *On Statistical Pattern Recognition in Independent Component Analysis Mixture Modelling*, Springer Theses, DOI: 10.1007/978-3-642-30752-2_8,

source extraction and signal classification are performed simultaneously, which could contribute to obtain higher insights into underlying physical phenomena from observations originated in real-world applications. Thus, this theoretical foundation was appropriate for dealing with the outlined problems. However, there were unsolved issues with the existing methods that we researched in this work: support of different kinds of learning; higher flexibility in the source modelling; support of different strategies for model parameter updating; and correction of residual dependencies. The formulation and testing of a general ICAMM framework, which solves these important open issues, represents a true challenge. It is particularly difficult if we consider the contexts of real applications where a priori knowledge of the data is incomplete; it is therefore complicated to derive a meaning for the parameters of the model (sources, bias terms, and mixing matrices).

Chapter 3 presented the first method of the research that addressed obtaining a general procedure by incorporating new features in ICAMM. This method attempts to obtain a balance between parametric and non-parametric estimation. Thus, the ICA mixtures were modelled by a short set of parameters maintaining simplicity; however, flexibility for source estimation with complex distributions was achieved by using a non-parametric kernel-based technique. The non-gaussianity of the data was preserved since any assumption about the source model was not imposed. The method allows unsupervised, supervised, and semi-supervised learning scenarios to be modelled in order to deal with different kinds of fragmented knowledge in specific applications. The advantages supplied by different kinds of ICA algorithms can be exploited since any ICA algorithm can be used in the model parameter updating. In addition, estimation of residual dependencies after training for correction of the posterior probability of every class to the testing observation vector was formulated. The capabilities of the method were demonstrated by means of an extensive number of simulations. Thus, ICA and ICA mixture data with different kinds of distributions such as Laplacian, uniform, Gaussian, Rayleigh, and K-type with several sample sizes were considered. The method was compared with standard ICA algorithms: InfoMax, Extended InfoMax, FastIca, JADE, and TDSEP as well as with non-parametric algorithms: Npica, radical, and Kernel-ICA. The results show competitive performance of the proposed method for accurate recovery of sources even for small sample sizes; improvement of the classification accuracy for data with non-linear dependencies using the proposed correction; and consistent learning from labelled-unlabelled data.

The second method of the research, explained in Chap. 4, consisted of an agglomerative hierarchical clustering procedure that creates higher levels of classification from a basic level of clusters formed by an ICA mixture model. This kind of organization allows an increasing degree of data model flexibility through hierarchical levels. Thus, the lack of flexibility of ICA projection models is compensated by the overall flexibility of the mixture of ICAs which in turn is relaxed by the complete hierarchy. An optimum cluster number is estimated using the partition and partition entropy coefficients. The use of hierarchy of relatively simple models offers greater ease of interpretation (different levels of generalization and abstraction) as well as the benefits of analytical and computational simplification. The application of the method was demonstrated for several simulations (compared with the single-linkage method) and the analyses of real images. In the simulations, the results showed that the binary trees obtained by the proposed method perform better than the single-linkage method to capture hierarchical data organization. Real data analysis showed meaningful dendrograms obtained for classification of images of objects with similar shapes, and segmentation of natural images.

The proposed methods build a versatile and powerful framework that can be employed in many real-world problems involving complex data densities. In Chaps. 5, 6, and 7, we provided several applications in signal classification and pattern recognition that processed different kinds of data: sonic and ultrasonic signals; EEG signals; images; and historic web log data. These chapters also included the following theoretical contributions: a method to introduce sequential dependencies in ICAMM-based classification (that was applied to sleep disorders analysis); and different approaches to establish the relation between the underlying physical model of the application and the probabilistic model ICA or ICAMM.

Besides the image processing analyses of Chap. 4, the application chapters dealt with the following challenging problems: diagnosis of the restoration of a historical building; classification of the chronological period of archaeological ceramics; classification of up to 12 kinds of different defective materials using impact-echo testing; detection of micro arousals in apnea patients; and detection of learning styles for the students in a virtual university (webmining). The ICAMM parameters enable a detailed explanation of the measured signals and their source data generators. In any case, even though the complexity of the problem constrains a physical interpretation, the framework can be used as a general data mining technique as demonstrated in the webmining application. The degrees of freedom afforded by the proposed methods allow the adaptation to and the solving of a broad range of real-world problems.

8.2 Contribution to Knowledge

This section lists the contributions that this thesis makes to the field of ICA and ICAMM. The thesis contributions are the following:

- A novel procedure for learning mixtures of independent component analyzers has been proposed (Sect. 3.3), which we call Mixca. The algorithm estimates the ICAMM parameters through the maximization of a single likelihood function. The technique of optimization applied was natural gradient, which simplifies the learning rule and speeds convergence. Four new extensions towards generalization of the ICAMM method were formulated: (i) incorporation of any kind of learning (unsupervised, semi-supervised, supervised); (ii) non-parametric kernel-based estimation of source densities; (iii) support

of any ICA algorithm for parameter updating; and (iv) correction of residual dependencies in classification after estimating the parameters of the model. Thus, the algorithm balances the parametric ICAMM formulation with increasing flexibility by using non-parametric source density estimation. Furthermore, requirements of prior knowledge (e.g., mixture proportions, density parameters, etc.) are not imposed, but all available priors -even fragmented knowledge- can be incorporated in the learning stage. This is an advantage in comparison with other methods such as Bayesian learning, which can require tuning of an extensive number of prior parameters.
The application of the proposed ICAMM technique has been demonstrated for several ICA mixtures and ICA datasets. The non-parametric approach of the procedure clearly yielded better results in source separation than standard ICA algorithms, indicating promising adaptive properties in learning source densities, even using small sample sizes. In addition, the estimation of multiple ICA parameters by the proposed method provides an important advantage of flexibility over other non-parametric and standard ICA algorithms when data with linear/nonlinear dependencies and complex structures are processed. The correction of the posterior probability has proven to be useful in the improvement of classification accuracy for ICA mixtures with nonlinear dependencies. The use of few parameters (such as the ICA model parameters) provides efficiency in the representation of the data mixture with the restriction of assuming linear dependencies in the latent variables of the generative data model. However, this restriction is relaxed by the proposed correction. Thus, the proposed method can be applied to a range of classification problems with data generated from underlying ICA models with residual dependencies, and it may be suitable for the analysis of real-world mixtures. In addition, the role of unlabelled data in training and classification for semi-supervised learning of ICA mixtures has been demonstrated, showing that unlabelled data can degrade the performance of the classifier when they do not fit the generative model of the data. The Mixca algorithm was exhaustively validated applying it in diverse simulations and real-world applications in the thesis. In addition, it was compared with standard classifiers (MLP, LDA), non-parametric ICA algorithms (Npica, Radical, Kernel-ICA), and standard ICA methods (InfoMax, extended InfoMax, JADE, TDSEP, FastIca).

- A novel method for agglomerative hierarchical clustering from mixtures of ICAs has been proposed (Sects. Sects. 4.2 and 4.3). The procedure includes two stages: learning the parameters of the mixtures (basis vectors and bias terms) using the Mixca algorithm and clustering the ICA mixtures following a bottom-up agglomerative scheme to construct a hierarchy for classification. The approach for the estimation of the source probability density function is non-parametric and the minimum Kullback–Leibler distance is used as a criterion for merging clusters at each level of the hierarchy.
The hierarchical clustering method was validated from several simulations and processing real data (image processing). Simulations showed the capability of

the method to generalize from close data densities and to detect outliers. This was compared with the traditional single linkage method that is based on distance between data objects. Image content similarity between objects based on ICA basis functions allowed an organization of objects in higher levels of abstraction to be learned (images of objects with similar shapes were grouped). Experiments with natural images showed the application to image segmentation based on the similarity of different patches of the image.

- A new procedure to incorporate sequential dependences in classification of ICA mixtures has been provided (Sect. 7.1). The so-called SICAMM method considers the case of having sequential dependence in the feature observation record. The algorithm is a sequential Bayes processor, which can be used to sequentially classify the input feature vector among a given set of possible classes. A hidden Markov model (HMM) was formulated using estimates of the class-transition probabilities and ICAMM parameters (mixture matrices, centroids, and source probability densities). These parameters were estimated using the proposed Mixca algorithm. Some simulations were presented to verify the improvement of SICAMM with respect to ICAMM. Moreover, a real data case was considered: the computation of hypnograms from apnea patients to help in the diagnosis of sleep disorders. Both simulated and real data analysis demonstrated the potential interest of including sequential dependence in the implementation of an ICAMM classifier.
- A pioneer application of the ICAMM methods for NDT using impact-echo technique has been presented in Sect. 5.3. The model is intended to defects with particular shapes or geometries, such as cracks, holes, and multiple defects. The model defined the determination of the quality condition of materials inspected by impact-echo (in homogeneous and different kinds of defective materials) as an ICA mixture problem. The model was formulated considering the impact-echo overall scenario as a MIMO-LTI system. A class of defective or homogeneous material was represented by an ICA model whose parameters were learned from the impact-echo signal spectrum. Thus, the resonance phenomenon involved in the impact-echo method was taken into account, and the compressed spectrum composed by contributions of every channel was formulated as observations for ICAMM. The ICA parameters of each class defined a kind of particular signature for the different defects.

  The proposed procedure was intended to exploit to the maximum the information obtained with the cost efficiency of only a single impact. To illustrate this capability, four levels of classification detail (material condition, kind of defect, defect orientation, and defect dimension) were defined, with the lowest level of detail having up to 12 classes. Results from extensive data sets from 3D finite element models and lab specimens of an aluminium alloy that contained defects of different shapes and sizes in different locations were obtained. The performance of the classification by ICA mixtures using Mixca was compared with LDA and MLP classification. We demonstrated that the mass spectra from impact-echo testing fit ICAMM, and we also showed the feasibility of this modelling to contribute to NDT applications.

- The accurate classification of archaeological ceramic fragments is quite a challenging application. Very often, shards provide no evidence (decoration, shape, etc.) about their origin, and they are obtained from deposits that are removed from their original manufacturing location. Therefore, we developed an ultrasound-based NDT method by applying ICAMM (implemented in the Mixca algorithm) for chronological cataloguing of archaeological ceramics. We applied Mixca in order to take advantage of its capabilities to deal with the arbitrary forms of the underlying probability densities in the feature vector space (even using small size training samples) and to take advantage of the possibility to incorporate partial labelling of the training samples, which can model the uncertainty of an expert in shard labelling. The features extracted from the ultrasonic signals were: centroid frequency; maximum frequency; bandwidth; maximum frequency amplitude; parameters A and β corresponding to an exponential model of the signal attenuation $\hat{x}(t) = Ae^{-\beta t}$; total signal power; propagation velocity; centroid frequency evaluated at time t_0; and higher-order statistics (time reversibility and third-order autocovariance). Mixca was tested using different variants depending on the embedded ICA algorithm (non-parametric, TDSEP, JADE, FastIca). The results demonstrated classification accuracy for shards from the following periods: the Bronze Age, Iberian, Roman, and the Middle Ages (from deposits in Eastern Spain). Physical interpretation of the results was provided considering various analyses: morphological and physiochemical characterization, ceramic composition and processing, and ultrasound propagation. The new method, which included a prototype, was patented. It is a real alternative to complement or replace destructive, costly, and time-consuming techniques, which are currently used by archaeologists for ceramic characterization.
- Novel applications in restoration of historical buildings were introduced. Two problems were approached: diagnosis of the consolidation status, and detection of layers in a wall of a heritage building. The ultrasound signals measured in auscultation of the wall were modelled as superposition of backscattering from the material microstructure plus sinusoidal waves. The sources recovered by Mixca (configured to estimate an ICA) allowed sinusoidal contributions to be accurately separated from the backscattering of the wall. The sinusoidal sources corresponded to non-consolidated zones and interferences. Therefore, enhanced B-Scans (images of the wall) were achieved using ICA as preprocessor to eliminate or separate the sinusoidal sources.
- Non-parametric ICA was introduced in webmining. The goal was to discover student learning styles from a huge historical web-log database of the use of facilities at a virtual campus. An ICA model was proposed assuming that the underlying independent sources that generated the web log data were

dimensions of the learning styles of the students. Thus, the observations were linear combinations of those styles through the use of the facilities at the virtual campus. Significant learning styles were detected for students of courses with grades. Lack of assessment in the courses did not allow learning styles to be identified. This confirmed a known pedagogical principle, that expected evaluation biases how the student learns.

8.3 Future Work

There are several open research topics that will improve the proposed methods. These topics are listed below:

8.3.1 Improvement of ICAMM Extensions

- Research on strategies to improve the parameter initialization. This is an important issue since the starting point of the process of learning determines the convergence of the algorithm. Regularization and penalization techniques can be applied to avoid convergence to local minima.
- Research on the residual dependence after convergence in classification and hierarchical clustering. The independence of the hidden variables is difficult to find in nature. Thus, information of class membership or posterior probabilities (i.e., the probability of every class conditioned to the feature of the observation vector) can be used to model fuzzy rules that reflect the dependency in the data mixture. In addition, dependency measures could be used to develop novelty detection procedures.
- Incorporation of the use of priors in the estimation of the source densities. This can be useful for particular applications in which features of the objective signals are known (statistical or spectral features such as bandwidth range, and kurtosis type). This is one of the current lines of research in ICA. However, the advantage of using priors to lead the algorithm to the objective is also a restriction that diminishes the blindness principle.
- Development of techniques to detect and process outliers. In some applications, the outliers are not strange data to be removed, but they are the interesting novelty to be found. The hierarchical classification of the data offers several possibilities to detect outliers such as the analysis of particular patterns in the manner in which the binary tree of clustering is built.

8.3.2 Extension of Mixca to Other Methods

- Extension to sequential methods. The proposed method uses block estimation of the ICAMM parameters whereas other methods employ sequential estimation of the parameters for each ICA. The extensions developed in this thesis can be used to generalize ICAMM sequential methods. The estimation of parameters such as stopping criterion should be improved since they are usually estimated arbitrarily in current techniques.
- Incorporation of on-line estimation of the parameters. The on-line learning processing would perform simultaneous structure and parameter identification. This kind of learning allows long processes to be monitored on-line. This assumes a dynamic modelling that can be supported by HMM (as we propose in this thesis) for incorporation of sequential dependencies in ICAMM.
- Development of a method for prediction based on ICAMM. This is an interesting issue that considers the strategy of linear local projections that can be adapted to partial segments of a data set while maintaining generalization given the mixture of several ICAs. The resulting algorithm could be applied to time series prediction, to recover missing data in images, etc.

8.3.3 Other Applications

- There is a myriad of possible applications where a modelling based on mixtures of ICAs can be valuable. For instance, event-related dynamics of brain oscillations to study sensorimotor processes, changes during the performance of cognitive tasks, or neurological disorders; development of query by visual example (QBVE) and query by semantic example (QBSE) systems; and to investigate in ICAMM-based physical models for NDT (associating features extracted from the NDT signals with physical properties of the materials).

Appendix
One-ICA Version of the Mixca Algorithm

ICA defines a generative model for the observed data. The conventional model is given by:

$$\mathbf{x} = \mathbf{A}\mathbf{s} \tag{A.1}$$

where $\mathbf{x}$ is the observed data as a random vector whose elements are the mixtures $x_1, \ldots, x_N$, $\mathbf{A}$ is a NxM matrix, called the mixing matrix, and $\mathbf{s}$ is a random vector with M elements $s_1, \ldots, s_M$, called the source elements.

The goal of ICA is to find a linear transformation of the data such that the latent variables are as statistically independent from each other as possible. Suppose y is an estimate of the sources, so that:

$$\mathbf{y} = \mathbf{W}\mathbf{x} \tag{A.2}$$

When the sources are exactly recovered, $\mathbf{W}$ is the inverse of $\mathbf{A}$.
The probability density function of the data $\mathbf{x}$ can be expressed as:

$$p(\mathbf{x}) = |\det \mathbf{W}|\, p(\mathbf{y}) \tag{A.3}$$

where $p(\mathbf{y})$ can be expressed as the product of the marginal distributions since it is the estimate of the independent components:

$$p(\mathbf{y}) = \prod_{i=1}^{M} p_i(y_i) \tag{A.4}$$

If we assume a non-parametric model for $p(y)$ we can estimate the source pdf's from a set of training samples obtained from the original dataset using equation (A.2). The marginal distribution of a reconstructed component is approximated as (kernel density estimation):

$$p(y_m) = a \cdot \sum_{n'} e^{-\frac{1}{2}\left(\frac{y_m - y_m^{(n')}}{h}\right)^2}, \quad m = 1 \ldots M \tag{A.5}$$

A. Salazar, *On Statistical Pattern Recognition in Independent Component Analysis Mixture Modelling,* Springer Theses, DOI: 10.1007/978-3-642-30752-2, where a is a normalization constant and h is a constant that adjusts the degree of smoothing of the estimated pdf. The learning algorithm can be derived using the maximum-likelihood estimation. In a probability context, it is usual to maximize the log-likelihood of equation (A.3) with respect to the unknown matrix $\mathbf{W}$:

$$\begin{aligned}\frac{\delta L(\mathbf{W})}{\delta \mathbf{W}} &= \frac{\delta \log|\det \mathbf{W}| \cdot p(\mathbf{y})}{\delta \mathbf{W}} \\ &= \frac{\delta \log|\det \mathbf{W}|}{\delta \mathbf{W}} + \frac{\delta \log p(\mathbf{y})}{\delta \mathbf{W}}.\end{aligned} \tag{A.6}$$

where

$$\frac{\delta \log|\det \mathbf{W}|}{\delta \mathbf{W}} = (\mathbf{W}^T)^{-1} \tag{A.7}$$

Imposing independence on y and using equation (A.4):

$$\begin{aligned}\frac{\delta \log p(\mathbf{y})}{\delta \mathbf{W}} &= \sum_{m=1}^{M} \frac{\delta \log p(y_m)}{\delta \mathbf{W}} = \sum_{m=1}^{M} \frac{1}{p(y_m)} \frac{\delta p(y_m)}{\delta \mathbf{W}} \\ &= \sum_{m=1}^{M} \frac{1}{p(y_m)} \frac{\delta p(y_m)}{\delta y_m} \frac{\delta y_m}{\delta \mathbf{W}}.\end{aligned} \tag{A.8}$$

where using equation (A.5):

$$\frac{\delta p(y_m)}{\delta y_m} = -a \sum_{n'} e^{-\frac{1}{2}\left(\frac{y_m - y_m^{(n')}}{h}\right)_2} \left(\frac{y_m - y_m^{(n')}}{h}\right) \frac{1}{h}, \quad m = 1 \ldots M \tag{A.9}$$

Let us call $\mathbf{w}_m^T$ the m-th row of $\mathbf{W}$. Then $y_m = \mathbf{w}_m^T \mathbf{x}$, and

$$\frac{\delta y_m}{\delta \mathbf{W}} = \mathbf{M}_m \tag{A.10}$$

where $\mathbf{M}_m(l, l') = \delta(l - m) x_{l'}$

Substituting equations (A.5), (A.9), and (A.10) in equation (A.8) we have:

$$\frac{\delta \log p(\mathbf{y})}{\delta \mathbf{W}} = \sum_{m=1}^{M} f(y_m) \mathbf{M}_m \tag{A.11}$$

where

$$f(y_m) = \frac{1}{h^2} \left[\frac{\sum_{n'} y_m \cdot e^{-\frac{1}{2}\left(\frac{y_m - y_m^{(n')}}{h}\right)_2}}{\sum_{n'} e^{-\frac{1}{2}\left(\frac{y_m - y_m^{(n')}}{h}\right)_2}} - y_m \right] \tag{A.12}$$

Considering the vector $\mathbf{f}(\mathbf{y}) = [f(y_1)f(y_2)\ldots f(y_M)]^T$, we can write

$$\frac{\delta \log p(\mathbf{y})}{\delta \mathbf{W}} = \mathbf{f}(\mathbf{y})\mathbf{x}^T \tag{A.13}$$

Using the results of equations (A.7) and (A.13), we may finally write equation (A.6) as:

$$\frac{\delta L(\mathbf{W})}{\delta \mathbf{W}} = (\mathbf{W}^T)^{-1} + \mathbf{f}(\mathbf{y})\mathbf{x}^T \tag{A.14}$$

Then we can apply equation (A.14) in the gradient updating algorithm to iteratively find the optimum matrix

$$\mathbf{W}(i+1) = \mathbf{W}(i) + a\frac{\delta L(\mathbf{W})}{\delta \mathbf{W}}(i) \tag{A.15}$$

Curriculum Vitae

Addisson Salazar received the B.Sc. and M.Sc. degrees in Information and Systems Engineering from Industrial University of Santander, the D.E.A. degree in Telecommunications from Polytechnic University of Valencia (UPV) in 2003, and the Dr. in Telecommunications degree from UPV in 2011. He also received the degrees of Senior System Analyst and Designer from Japan International Cooperation Agency in 1996 and Expert in Information and Telecommunication Technologies from University of Alcalá in 2008. He obtained the Accreditation to Associate Professor from the Valencia Commission of Accreditation and Quality Evaluation in 2005. Since 2002, he has been with the Signal Processing Group in the Institute of Telecommunications and Multimedia Applications at UPV where he was appointed as an official scientific staff member in 2007. His research interests include statistical signal processing, machine learning, and pattern recognition with emphasis on methods for signal classification based on time-frequency techniques, blind source separation and mixtures of independent component analyzers. The application of his research has been focused on non-destructive testing and biomedical problems.

A. Salazar, *On Statistical Pattern Recognition in Independent Component Analysis Mixture Modelling*, Springer Theses, DOI: 10.1007/978-3-642-30752-2,

MIX
Papier aus verantwortungsvollen Quellen
Paper from responsible sources
FSC® C105338

If you have any concerns about our products,
you can contact us on
ProductSafety@springernature.com

In case Publisher is established outside the EU,
the EU authorized representative is:
Springer Nature Customer Service Center GmbH
Europaplatz 3, 69115 Heidelberg, Germany

Printed by Libri Plureos GmbH
in Hamburg, Germany